The EAAP series is published under the direction of Dr. P. Rafai

EAAP – European Federation of Animal Science

The European Association for Animal Production wishes to express its appreciation to the *Ministero per le Politiche Agricole e Forestali* and the *Associazione Italiana Allevatori* for their valuable support of its activities

Evaluation of carcass and meat quality in cattle and sheep

Evaluation of carcass and meat quality in cattle and sheep

EAAP publication No. 123

Editors:

C. Lazzaroni, S. Gigli and D. Gabiña

Wageningen Academic
Publishers

ISBN 978-90-8686-022-7

ISSN 0071-2477

First published, 2007

Wageningen Academic Publishers
The Netherlands, 2007

Table of contents

Introduction

Sergio Gigli[1] *and Dunixi Gabiña*[2]
[1]*C.R.A. Istituto Sperimentale per la Zootecnia, Via Salaria 31, 00016 Monterotondo Scalo (Roma), Italy*
[2]*CIHEAM-IAMZ Mediterranean Agronomic Institute of Zaragoza, Apartado 202, 50080 Zaragoza, Spain*

Livestock production systems will only be sustained in the long term if their products continue to meet the demand of consumers. The quality of ruminant carcasses, meat and meat products is of predominant importance in a competitive market where consumers tend to have a formed opinion about the criteria that define meat quality such as flavour, tenderness, juiciness, smell, colour and texture. The carcass evaluation could be interesting as a precocious classification of the final quality of meat coming from each carcass.

Today the quality characteristics of the meat must be different according to its utilisation (supermarket, butcher, catering, refectory, etc.) and so it is very important to choose very early the final destination of the carcass. Obviously the carcass classification must be correlated with meat quality characteristics required by final consumer. Other important factors that have to be taken into account in order to maintain a demand for ruminant meat are safety and traceability.

This book is originated in the session on 'New developments in evaluation of carcass and meat quality in cattle and sheep' organised in the 57th Annual Meeting of the European Association for Animal Production (EAAP) held in Antalya (Turkey) on 17-20 September 2006 and compiles 23 contributions to the session. The objective of the session was to review the historical and recent developments for carcass evaluation and grading and for meat quality assessment in beef and sheep placing special emphasis on new concepts and approaches to define carcass and meat quality and on the use of modern technologies for composition and quality evaluation. A range of technologies have been presented such as ultrasounds and colour reflectance, X-ray computerised tomography, spectral and thermal imaging and image analysis and NIRS. The use of phenotypic markers such as the plasma hormones and genetic markers to predict carcass composition and meat quality has also been presented.

Evaluation of carcass and meat quality in the EAAP

Carla Lazzaroni
Dipartimento di Scienze Zootecniche, Università di Torino, via L. da Vinci 44, 10095 Grugliasco (Torino), Italy

Abstract

Evaluation of carcass and meat quality was carried out since the meat market started and the consumers become more conscious about the possible differences among carcass and meat. The developing of technology allowed researchers to better evaluate the product characteristics as requested by consumers, and a survey of such trend is here outlined referred to EAAP.

Keywords: carcass evaluation, meat quality, EAAP activities

Introduction

Evaluation of carcass and meat quality, in all species of livestock but especially in cattle and sheep, is a very important topic in the meat marketing, since it left a local perspective for a national and international one. This happened notably in the 20th century, due to a massive urbanisation of population and to the large gap between the productive world (farmers, breeders, etc.) and the consumers, also supported by the improved methods of meat conservation, as refrigeration and freezing. So it become more and more important to be able to describe the characteristics of the product to be sold – both carcass and meat – in a proper and unique way, understandable also for people without any direct experience of farming and livestock.

But what does it mean carcass and meat quality? How can we indicate carcass – and meat – quality? Is such idea of quality the same all over the world? Such topics will be discussed in the following papers, but few concepts should be clear:

- A wide range of animal is used for meat production, different for breed, rearing system, gender, age (from 6 months to more than 6 years), live weight, etc., and all those characteristics affect carcass and meat quality.
- Each animal is nearly a unique ensemble of a great number of genes, and we don't know enough their effects on meat production.
- For the consumer it could be very interesting find on the label of the purchased meat some indication on meat quality (tenderness, juiciness, fatness, etc.) but such characteristics are heavily affected by meat preparation at home (cooking technique and time, temperature, use of dressing, etc.), so it become quite difficult give such information referred to the dish we can found on our table, also because the consumers often don't have a clear idea of the proper way to cook and prepare all different cuts of meat.

EAAP activities on carcass and meat quality

Inside the European Association for Animal Production the topic of carcass evaluation and meat quality was present since 1965 (see 'EAAP News' on *Livestock Production Science*). In fact, connected to the 8th Study Meeting held in Noordwijk (The Netherlands) in 1965, a Symposium on 'Carcass value' took place and in the same year a Working Group on 'Carcass evaluation and meat quality' was created with members of different Study Commissions.

The Working Group developed several activities, and in 1968, during the 9th Study Meeting of EAAP held in Dublin (Ireland), a Symposium on 'Methods for carcass evaluation' was organised. In 1971, the Working Group was transformed to the Working Party on 'Assessment of carcass characteristics in cattle' (H. De Boer, NL; B.L. Dumont, F; R.W. Pomeroy, UK; J.H. Weniger, D), and associated with the Study Commission on Cattle Production.

The first result of the Working Party was the publication of the 'Manual on E.A.A.P. reference methods for the assessment of carcass characteristics in cattle' (De Boer *et al.*, 1974), and such methodology was later taken up by the European international bodies involved in beef classification and marketing (Council Reg. EEC 1208/81).

In the same year 1974 a Questionnaire on 'Cattle carcass dissection techniques' was sent to several research units all over Europe (36 institutes in 20 countries) to evaluate and classify the different methods for beef carcass dissection methods, as the main aim of the Working Party was to promote understanding between scientists of different countries. A report on the variety of methods used for dressing, measuring, jointing and tissue separation in beef animals was presented to the Commission on Cattle Production by R.W. Pomeroy (UK) during the EAAP Meeting held in Copenhagen (Denmark) and the results were discussed in a EEC Seminar on 'Criteria and methods for the assessment of carcass and meat characteristics in beef production experiments', held in Zeist (The Netherlands) in 1975 (Fisher *et al.*, 1976).

The collaboration of EAAP members with the EEC Programme of Co-ordination of Research on Beef Production continued in the following years, and a CEC Working Sub-Group on 'Carcass and meat quality' was established. The activities of the Working Sub-Group included some meetings on 'Standardisation of criteria for assessing meat quality' (1976-1977) and a training course on 'Beef carcass classification', held in 1977 in Langford (UK), and some reference publications (Meat Research Institute, 1976; Williams and Bergstrom, 1976; Brown *et al.*, 1978; Speight, 1979).

In the same years, during the EAAP Meeting held in 1975 in Warsaw (Poland) the attention was drown also to sheep, and a report on 'Sheep carcass grading and marketing of lamb' was presented to the Commission on Sheep and Goat Production by R. Wassmuth (D), followed during the EAAP Meeting held in 1980 in Munich (Germany) by a session on 'Quality in lamb and kid carcasses' and 'Practical demonstration of lamb and kid carcasses judgement', both organised by F. Colomer-Rocher (E).

Finally, also pig carcasses were studied in a session on 'Pig grading for carcass quality', organised by B. Desmoulin (F) during the EAAP Meeting held in 1981 in Zagreb (Yugoslavia).

In the meantime, the activity of EAAP members connected to the Commission of the European Communities on 'Beef Production Research Programme' continued: the 'Procedures for measuring meat quality characteristics in beef production experiments' were published (Boccard *et al.*, 1981), and the possibility to estimate body composition and retail value of carcass using non invasive techniques were investigated in several workshop and meetings (Bech Andersen, 1982; Bech Andersen *et al.*, 1982; Fisher, 1983; Lister, 1984).

Such activities were of topical interest in the Eighties, and were often discussed also inside the EAAP: during the Meeting held in 1983 in Madrid (Spain), in a session on 'Beef carcass classification: methods and application' organised by H. De Boer (NL); during a Symposium on 'Carcass grading' held in 1984 in The Hague (The Netherlands); during the Meeting held in 1985 in Thessaloniki (Greece), in a session on 'Evaluation of carcass quality in live pigs' organised by A.J. Kempster (UK);

during the Meeting held in 1986 in Budapest (Hungary), in a session on 'Evaluation of live animals and carcasses for meat production (sheep)' organised by P. Morand Fehr (F) and P. Susmel (I).

In Budapest was also established a new EAAP Working Group on 'Carcass evaluation', chaired by H. De Boer (NL) and collecting 35 carcass and meat experts from 16 countries, and which activities aroused also the interest of the Division for Coordination of Agricultural Research of the European Commission. The priority items to be examined were identified as: instrumental pig carcass classification and prediction of lean content; reference methods for carcass assessment in lamb and sheep; non-invasive techniques for carcass and live evaluation in farm animals; beef carcass characteristics for selection; qualitative characteristics of lean and fat.

In the following years the Working Group suffered of period of inactivity due to the lack of financial support for the project on sheep carcass assessment, and in 1991 it was decided to finish the current activities on sheep carcass assessment and on terminology in the forthcoming year and to present it in the session on 'Body composition of the live small ruminant: methods of assessment and their application' organised by A. Kempster (UK) during the EAAP Meeting to be held in 1992 in Madrid (Spain); to establish a new working party on non-invasive techniques for carcass and live assessment; and to decide in Madrid whether the original broad working group should be continued or replaced by more specific working groups.

Nevertheless in the same years the interest in carcass evaluation continued, as testified by several workshop and symposia as: the CIEHAM-AGRIMED Workshop 'Les carcasses d'agneaux et de chevreaux méditerranéens' held 1986 in Zaragoza (Spain) (Flamand and Gabiña, 1988); the EAAP Symposium on 'New techniques in pig carcass evaluation' held in 1988 in Helsinki (Finland) (O'Grady, 1988); the EC Programme for Coordinated Agricultural Research Seminar 'Application of NMR techniques on the body composition of live animals' held in 1988 in Mariensee (Germany) (Kallweit *et al.*, 1989); the EAAP Symposium on 'Beef carcass and meat quality evaluation' held in 1991 in Dummerstorf – Rostock (Germany) (Ender, 1991); the CIEHAM Courses 'Carcass and meat quality in ruminants' held in 1991 and 1994 in Zaragoza and in 1992 in Valencia (Spain).

The project on 'International Reference Procedures for Sheep Carcasses' of the EAAP Working Group on 'Carcass evaluation' was concluded in a joint meeting held in Zaragoza (Spain), when 25 experts from European and Mediterranean countries produced a basic report as a general reference on agreed measurements and dissection procedures (Fisher and De Boer, 1994). In the meantime it was decided to dissolve the Working Group and to replace it with *ad hoc* groups, if needed, on topics of current interest to EAAP participants, as the Working Group on 'Non-invasive techniques of carcass assessment', coordinated by E. Kallweit (D).

The Working Group on 'Non-invasive techniques of carcass assessment', coordinated by E. Kallweit (D), originated during the EAAP Meeting held in 1993 in Aarhus (Denmark), following the session on 'Improved methods of carcass and meat quality estimation', as with the EEC Common Market more research seems to be needed to improve the standardisation of carcass grading. About 20 experts of 9 different countries were invited to joint the Working Group to work on the following topics: ultrasound, velocity of ultrasound, total body electrical conductivity (TOBEC), bioelectrical impedance analysis, X-ray CT, magnetic resonance imaging (MRI), magnetic resonance spectroscopy (MRS), video image analysis (VIA), laser technique, microwave imaging, comparison for practical application and economical aspects.

A report on the activity of the Working Group was presented by E Kallweitt (D) during the EAAP Meeting held in 1994 in Edinburgh (United Kingdom), followed by an expert meeting held in 1995 in

Kaposvar (Hungary), were participants were asked to present an updated survey about his particular field of interest. Further activities were again stopped due to the lack of financial support.

The evaluation of carcass and meat quality inside the EAAP seems to be not so interesting during the following ten years, characterised by environmental and productive topics, so it looks interesting and necessary to have a session on 'Indicators of milk and beef quality' during the EAAP Meeting held in 2004 in Bled (Slovenia) (Hocquette and Gigli, 2005), followed by a session on 'New developments in evaluation of carcass and meat quality in cattle and sheep' during the EAAP Meeting held in 2006 in Antalya (Turkey), the last one to have an update on topics as: developing of carcass evaluation and grading for the European and non-European markets; future developing of carcass evaluation and grading; link between carcass and meat evaluation; today situation for carcass and meat evaluation in the different countries; etc.

References

Bech Andersen, B. (ed.), 1982. In vivo estimation of body composition in beef. Landhusholdnings Selskabets Forlag, Rolighedsvej, Copenhagen, Denmark, 195 pp.

Bech Andersen, B., H. Busk, J.P. Cladwick, A. Cuthbertson, G.A.J. Fursey, D.W. Jones, P. Lewin, C.A. Miles and M.G. Owen, 1982. Ultrasonic techniques for describing carcass characteristics in live cattle. Commission of the European Communities, Luxembourg, Publication No. EUR 7640, 78 pp.

Boccard, R., L. Buchter, E. Casteels, E. Cosentino, E. Dransfield, D.E. Hood, R.L. Joseph, D.B. MacDougall, D.N. Rhodes, I. Schön, B.J. Tinbergen and C. Tourraille, 1981. Procedures for measuring meat quality characteristics in beef production experiments. Report of a Working Group in the Commission of the European Communities' (CEC) Beef Production Research Programme. Livestock Production Science, Volume 8, 385-397.

Brown, A.J., H.E. Coates and B.S. Speight, 1978. A photographic guide to muscular and skeletal anatomy of the beef carcass. Meat Research Institute, Langford, UK, 186 pp.

Council Regulation (EEC) No. 1208/81. Community scale for the classification of carcases of adult bovine animals. Official Journal of the European Communities No. L 123, 7 May 1981.

De Boer, H., B.L. Dumont, R.W. Pomeroy and J.H. Weniger, 1974. Manual on E.A.A.P. reference methods for the assessment of carcass characteristics in cattle. Livestock Production Science, Volume 1, 151-164.

Ender, K. (ed.), 1991. Beef carcass and meat quality evaluation. Dummerstorf – Rostock, Germany, 165 pp.

Fisher, A.V. (ed.), 1983. Comparative retail value of beef carcasses. Commission of the European Communities, Luxembourg, Publication No. EUR 8465, 167 pp.

Fisher, A.V. and H. De Boer, 1994. The EAAP standard method of sheep carcass assessment. Carcass measurements and dissection procedures report of the EAAP working group on carcass evaluation, in cooperation with the CIHEAM Instituto Agronómico Mediterráneo de Zaragoza and the CEC Directorate General for Agriculture in Brussels. Livestock Production Science, Volume 38, 149-159.

Fisher, A.V., J.C. Tayler, H. De Boer and D.H. Van Adrichem Boogaert (eds.), 1976. Criteria and methods for the assessment of carcass and meat characteristics in beef production experiments. Commission of the European Communities, Luxembourg, Publication No. EUR 5489, 406 pp.

Flamand, J.C. and D. Gabiña (eds.), 1988. Les carcasses d'agneaux et de chevreaux méditerranéens. OOPEC, Luxemburg, Publication No. EUR 11479, 165 pp.

Hocquette, J.F. and S. Gigli (eds.), 2005. Indicators of milk and beef quality. Wageningen Academic Publishers, Wageningen, The Netherlands, EAAP Publication Series No. 112, 464 pp.

Kallweit, E., M. Henning and E. Groeneveld (eds.), 1989. Application of NMR techniques on the body composition of live animals. Elsevier Applied Science Publishers, London, UK, Publication No. EUR 11713, 174 pp.

Lister, D. (ed.), 1984. In vivo measurement of body composition in meat animals. Elsevier Applied Science Publishers, London, UK, 241 pp.

Meat Research Institute (ed.), 1976. Beef carcasses: methods of dressing, measuring, jointing and tissue separation. EAAP Publication Series No. 18, 79 pp.

O'Grady, J.F. (ed.), 1988. New Techniques in pig carcass evaluation. Pudoc, Wageningen, The Netherlands, EAAP Publication Series No. 41, 105 pp.
Speight, B.S., 1979. Standardised photography of beef carcasses. Meat Research Institute, Langford, UK, 36 pp.
Williams, D.R. and P.L. Bergstrom, 1976. Anatomical jointing, tissue separation and weight recording proposed as the EEC standard method for beef. Commission of the European Communities, Luxembourg, Publication No. EUR 5720, 27 pp.

Beef carcass classification in the EU: an historical perspective

Alan Fisher
University of Bristol, Division of Farm Animal Science, Department of Clinical Veterinary Science, Langford, Bristol BS40 5DU, United Kingdom

Abstract

Beef carcass classification plays an important role in Europe, as a marketing aid within and between countries and as a means of increasing the precision of price reporting for administrative purposes. This paper reviews differences between classification and grading and how this influenced the thinking in the development of new schemes. Some experimental evidence from published work is included to show how the ability of assessors to discriminate consistently between carcasses is dependent on the nature of the characteristic evaluated. The options of sorting carcasses according to meat quality or yield are discussed, and some of the difficulties in gaining acceptance of national schemes by the meat industry are presented. A brief account of schemes used in Europe prior to harmonization is presented, together with the role of the EAAP in the evolution of the pan-European scheme culminating in the current EUROP system.

Keywords: carcass grading, fatness, conformation, harmonization, EUROP

Introduction

Cattle reared in temperate climates are of a type, distinct from the humped cattle of the tropics which previously were considered to be a different species. But even within the temperate zone, cattle are inherently very variable, their obvious phenotypic variation encompassing size, shape, and coat colour. This is primarily due to the variety of breeds that have originated, although gender and husbandry are also important factors. These, plus variation in age at slaughter, confer much variability on their carcasses which not only vary in size and shape but also, quite obviously, in fatness. So, when animals or carcasses began to be traded for meat (as opposed to being slaughtered and consumed at home), there was a need to discriminate between them. The earliest attempts at categorizing thus had their roots in trading many years ago and even formalised schemes have been used in some countries of the world for a long time; for example, the first tentative United States standards for grades of dressed beef were formulated in 1916 (United States Department of Agriculture, 1997). However, it is only in the last fifty years or so that more widespread developments, based on new approaches to sorting carcasses, have been made.

This review examines the methods that have been used to categorize beef carcasses, the rationale for their deployment and the evolution of schemes leading to those used today. The emphasis is on Europe but some of the early, and notable, developments were in the USA and exporting countries and they cannot be ignored in any historical commentary. Furthemore, there is, inevitably, a strong UK perspective as the author was able to trace documented material, some rather obscure, in his home country, a process that would have been very difficult to undertake in other countries, and the Meat and Livestock Commission, or key members of its staff, are cited frequently because of the important role that organisation played, both nationally and at the European level.

Carcass grading or classification? For meat quality or meat yield?

One of the earlier carcass descriptive schemes that took on national significance was the USDA beef grading in the USA. A uniform sorting system was seen as advantageous in the marketing of beef carcasses but the emphasis was on indices of meat eating quality rather than on carcass quality, the pressure coming from supporters of the traditional breeds, such as Angus and Hereford, who believed their stock provided superior quality beef. Yield grades for beef in the USA, as an adjunct, or alternative, to the quality grades, were not brought in until much later (1969). This contrasts with the situation in Europe where direct meat quality assessment has rarely featured in commercial schemes. This is not because meat quality is not important in Europe but it does reflect, to some extent, the degree of acceptance that the eating quality of meat can be determined by the type of assessment or measurement that can easily be made on the carcass at, or within a day or so of, slaughter. Also, it probably reflects the range, and distribution within that range, of the supposed indicators of quality. For example, intramuscular fat or marbling is at a higher level in the cattle population of the USA than that of Europe, with many more examples at the top end of the range.

In the USDA quality grading, the degree of marbling and the maturity of the animal are key features on which carcasses are sorted. Such allocations are on the basis of an assumed best/worst quality and the names given to the categories reflect this. Thus the top quality is Prime (high level of marbling, youthful), followed by Choice, Good, Standard (progressively less marbling fat) and then, in the more mature category, Commercial (high level of marbling), followed by Utility and Cutter (progressively less marbling). But a carcass could fall within a particular category based on a number of permutations of maturity and marbling. For example, to qualify as Choice, a carcass could be maturity A (most youthful sector) and marbling 'Small' (middle of the range), or maturity B (less youthful) and marbling 'Slightly abundant' (top of the range). This placing of carcasses into categories having preconceived values, but in which carcasses within a tier may vary for a variety of phenotypic reasons, is known as grading.

At first sight, it may seem that the concept of grading beef for palatability is universally sound. Surely all sectors of the industry would prefer to utilize and promote meat that is tender rather than tough? But for certain manufactured products it may be preferable to process meat that does not disintegrate too easily, and if the fat content can vary within a grade this is not helpful for processors wanting to produce a consistent, or specified low fat, product. So grading for assumed meat quality can have its drawbacks and grading carcasses for the rather more overt, physical characteristics, such as fatness or size, is even more confounding. Although such grading was used in countries which exported large volumes of meat and it fulfilled some of the aims of early export schemes (where only carcasses 'making the grade' were exported), grading was not suitable for general commercial carcass marketing in which different kinds of carcasses were required for different sectors of the industry. In the late 1960s, evidence was beginning to emerge that simplified carcass descriptors were more objective than grading (e.g. Williams, 1969). This is discussed in detail below, but it is important to note that there was, at this time, a gathering body of opinion that the way forward was to classify carcasses on the basis of important economic characteristics (but whose evaluation was on a scale that itself was free of imputed value), and this was to be preferred to grading.

The classification of objects, namely the allocation of items to groups (classes) in which items are similar, but different from those in other groups, appears to be an important element in human cognitive development (proposed, among others, by Piaget, the developmental psychologist). It therefore 'comes naturally' to classify carcasses based on their appearance and this simple approach, based on visual assessment, has been the basis of a system that is still used in commercial beef trading today. This concept of classification is founded on a descriptive process and a key feature in its success is the use of a common language to describe characteristics of carcasses that have commercial significance. The distinction between grading and classification was repeatedly emphasised by researchers and industry analysts in the 1970s (Harrington, 1973b; Pomeroy, 1977) and, more recently by Price (1995) in a review of developments in classification and grading. In Europe, at least, the message has been heeded and the official scheme for beef is one of classification and not grading. It is important, though, to recognize that when two or more carcass attributes are classified, and their ratings are subsequently combined to sort carcasses, this may be seen as grading. Indeed, this is how classification information is often used within individual meat companies (to define their requirements in a specification), but in no way does this weaken the case for classification which always conveys the unambiguous, primary data, enabling that specification to be drawn up.

More recently, it can be seen that 'grading' is becoming more commonly used in the carcass context. Perhaps the distinction from classification is seen as pedantic and the two terms can be used synonymously. But consider the sorting of pig carcasses: this was based originally on fat thickness and carcass weight – both being classified objectively (some schemes also included a visual assessment of conformation) and sellers and buyers had clear descriptors to work on. In the official EC pig carcass scheme (Commission of the European Communities, 1984), combinations of fat (and muscle) depths, plus carcass weight, are used to predict percentage lean in the carcass which is then allocated to a 'grade' (coded S, E, U, R, O, P where S contains 60% or more lean, P has less than 40%). This allocation suffers from one of the drawbacks noted earlier, i.e. the percentage lean is based on permutations of different values for the key measurements and unambiguous information for traders is obscured. This is why there has been a reluctance, by meat traders in some countries, to use predicted lean percentage and they continue to base negotiations on fat thickness. The term 'grading' is also embodied, rather confusingly, in the codified version of the EC Council regulation determining the community scale for the classification of carcases of adult bovine animals (CEU, 2006): 'A Community *grading* scale for the *classification* of carcases of adult bovine animals should be applied for recording prices and for intervention in the beef and veal sector'.

An extreme example of grading, based on classified and graded primary data combined to give an overall result, is embodied in the carcass competition score card. In carcass competitions, held at agricultural and meat industry shows and exhibitions, a number of carcass characteristics are awarded scores in order to have a structured basis for awarding prizes (Starke and Joubert, 1961). Some of these carcass attributes may be objective (for example, fat and muscle depths on a cross section of the loin) whilst others can be sensory (for example, 'legs short boned and well filled'). Traits are weighted differently, based on their preconceived economic importance and a notion of the 'ideal carcass', so they have different maximum scores (for example, 10 for fat depth and 20 for each of muscle depth and leg shape, in the example above). The sum of the scores for the different traits is interpreted as a measure of overall carcass quality and the 'best' carcass is the one having the highest score.

Williams (1972) conducted an assessment trial based on the score card to show which of the many carcass characteristics that can be assessed are done so consistently and reliably. Six very experienced judges assessed 21 beef carcasses using a score card (Table 1) and agreement between them was determined as the coefficient of concordance (W) for each of the attributes assessed. The higher the value of W, the better the agreement but as Table 1 shows, agreement was not high for any of the attributes and the judges differed more widely in their assessment of what constitutes ideal fat cover than in their assessment of muscular development. So, as a next step in the investigation into the reliability of visual assessment, Williams broke down the complex quality attributes in the score card into their component parts so that only a single characteristic was assessed at any one time. He used seven point scales and provided a written description of the classes and, importantly, none of the 'new' assessment scales had the concept of an optimum inherent in them. The trial used lamb carcasses but the principle was exactly the same for beef. For example, the traditional score card for lamb had the following as an item for scoring: 'legs – short and well-filled' (20 marks). In the

Table 1. Coefficients of concordance (W) between six experienced judges in the ranking of 21 beef carcasses for 9 characteristics.

	Maximum mark	W
1. Lightness of neck and rib and adequate depth of shoulder flesh	10	0.40
2. Fat cover on forequarter not excessive	5	0.41
3. Fat cover on hindquarter even and adequate	10	0.42
4. Shin, muscle and bone proportion	5	0.49
5. Depth of fat over 'eye' muscle	10	0.50
6. Kidney, channel, cod or udder fat not excessive	10	0.56
7. Depth of 'eye' muscle	10	0.67
8. Quality, texture, marbling and colour of 'eye' muscle	20	0.69
9. Muscular development of hindquarter	20	0.69
Total	100	

revised system this was broken down into length of leg (1=extremely short; 2=very short; 3=short; 4=average; 5=long; 6=very long; 7=extremely long) and plumpness of leg (1=extremely poorly developed; 2=very poorly developed; 3=poorly developed; 4=average; 5=well developed; 6=very well developed; 7=extremely well developed). Eight judges of varying experience assessed 15 carcasses on four occasions with the carcass identification numbers being changed at random after each assessment occasion. The results, expressed as variance ratios from an analysis of variance (variation between carcasses: variation between repeat assessments) showed conclusively that the revised attributes were assessed more discriminantly and with better repeatability than the traditional score card characteristics.

The importance of sorting carcasses

The underlying principle of carcass classification is that a description of each carcass, at the end of the slaughter line, can improve the efficiency of the meat industry by facilitating decisions on a number of animal production and marketing issues. Of course, the overall benefits must exceed the costs of execution and to achieve its full potential, the percentage of national classifications must be high, the classifications must be commercially relevant, practicable, reliable, and comparable between abattoirs (Meat & Livestock Commission, 1975).

There are several specific reasons for the deployment of a classification system and the advantages to be gained that can be listed but the five main advantages are given below.

1. To provide the farmer with information on the requirements of retailers and, ultimately, consumers, so that he can more precisely define his short term marketing decisions and his longer term production decisions.
2. To enable carcass purchasers to specify their needs and reduce variability.
3. To provide a basis for pricing; producer-wholesaler and wholesaler-retailer transactions.
4. To enable remote buying and therefore reduce costs.
5. To aid the precision of price reporting to the European Commission and thus contribute to the intensification of competition.

In addition, for some regions or countries of the world, because of their main trading channels or because the way the scheme has developed, two others can be added.

6. To aid exports by controlling quality and by promotional schemes.
7. To provide information that can be conveyed to the consumer, strengthening demand and helping promotional drives.

Point (6) was important in exporting countries like Australia, Argentina and New Zealand whilst (7) became a feature of the USDA quality grading scheme which acquired promotional significance in the USA. However, in Europe it was objectives (1) to (5) that were the drivers but these were far less attractive to the meat industry. This aspect - the need to convince the industry of the benefits of a national (or, later, a European) scheme - has probably become obscured with the passage of time but it was a real issue in the 1960s and 1970s when there was much discussion about the adoption and implementation of carcass descriptive schemes. The advantages of having a common language to describe carcasses, even within a country, to improve marketing efficiency and transparency, seemed clear to academics but was resisted by many meat traders, perhaps because of concerns that the industry would have to bear operational costs.

This resistance to a national classification system occurred at different levels of the meat chain. Farmers were traditionally paid on a flat-rate basis for a batch of animals sent for slaughter; over-fat or poor conformation animals received the same price per kilogram as leaner, better conformed individuals. Payment based on a carcass classification meant that farmers who produced below average animals would be penalised and there has always been a degree of mistrust in a subjective visual assessment. At the wholesale level, buyers using their own personal skills to select 'better' animals were antagonistic to an introduced system that deprived them of this power to make important decisions. Also, there was concern that retailers would use the classification 'language' to specify a very narrow range of acceptable carcasses and wholesalers would find it difficult to meet this demand whilst, at the same time, be faced with the problem of disposing of carcasses falling outside the mainstream.

Methodology

In the USDA Quality and Yield Grading schemes, carcasses are assessed after chilling and some of the measurements used to define the grades are made on the exposed cross-section of the carcass, at a specified plane, following ribbing (the partial separation of fore and hindquarters exposing the tissues close to the vertebral column). The tissue areas and thicknesses obtainable from such a cross-section are useful predictors of the composition of the carcass and lend themselves to objective measurement. However, this approach was unusual as most grading schemes were based on assessment at the end of the slaughter line, on the hot carcass or sides, and this was true in the UK, even though the former 'might have been preferred' (Harrington, 1973a). It appears that there were two reasons for this: in the UK, there was already in existence a subsidy payment (known as the Fatstock Guarantee, which continued until 1973) based on the certification of carcasses and, later, on weight authentication, and the new classification had to link in with these practices that were carried out at the carcass weigh point (otherwise carcasses would have to be examined twice); second, it was not trade practice generally throughout Europe, at the time that classification was being introduced, to quarter carcasses at a set time after slaughter and many abattoirs only quartered sides immediately before loading on to the removal lorry. This, compounded with local variation in the quartering plane and the early departure (before 24 hours) of sides from some abattoirs, restricted the options for classification.

Interestingly, the later-implemented Japanese (JMGA, 1988) and Australian beef Grading schemes (Meat Standards Australia, undated) follow the USDA example of measuring the cold, quartered sides.

The thinking that led to revised national schemes and then to an EC scheme, was that the classes should have relevance for the economics of cutting and selling the carcass and aspects of 'quality' that could not demonstrably be linked to realisation value should not be included. Harrington (1973a) identified three factors that influence the profitability to a retailer from cutting and selling a carcass (over and above the cutting method used and the buying and selling prices). These three factors are the characteristics associated with meat quality, the quantity of saleable meat and the distribution of meat in the carcass across cuts having different unit values. These are shown in Table 2, together with the characteristics influencing each of the factors and how they may be accommodated in a classification scheme applied to hot carcasses or intact sides at the end of the slaughter line.

Some of the possible components listed in Table 2 can only be assessed or measured in a limited way in the intact side, for example colour of lean. This would only be possible on the muscles exposed when the carcass was split, namely the muscles lying between the dorsal processes of the vertebrae or those lying distally to the symphysis pelvis (aitch bone). At the time of assessment, the muscles between the vertebral processes will not be oxygenated because the carcass will just have been split so they will have the purple colour of myoglobin; the muscles over the symphysis pelvis (mainly *adductor*) will have been exposed to the air for longer (dependent on the line speed) but may be partially oxygenated and thus exhibiting the increased redness of oxymyoglobin. Such variation may invalidate the inclusion of lean colour in a classification scheme and a decision on which traits to include can be made on the basis of similar critical analyses. As far as the Meat & Livestock Commission were concerned, they selected for inclusion those components shown in bold

Table 2. Factors affecting beef carcass realisation value and their possibly indicative traits that could be incorporated in a classification scheme (after Harrington, 1973a).

Factors	Characteristics	Source of variation in characteristics	Possible components of a classification scheme
Meat quality	Tenderness Flavour Juiciness Water-holding capacity Colour of lean Colour of fat	Pre-and post-slaughter treatment Growth rate Feed Breed Gender Maturity Fatness	*Age (dentition)* *Weight* *Gender* Breed *Fatness* (thickness or *visual*) Colour of fat (instrumental or visual) Colour of lean (instrumental or visual) Texture of lean (visual)
Amount of saleable meat	Yield of saleable meat	Fat trim (subcutaneous, intermuscular) Bone yield Other tissues trim (blood vessels, nerves, fasciae, tendons, bruised/bloody muscle, etc.)	*Weight* *Gender* Breed *Fatness* (thickness or *visual*) *Conformation* Muscle depth
Distribution of meat in carcass	High priced cuts % Shape of cuts	Lean meat distribution Fat distribution Muscle thickness	Breed *Fatness* (thickness or *visual*) *Conformation* Weight per unit length Other 'shape' measurements Muscle depth

in Table 2. The arguments for and against the different components is given in detail in the paper by Harrington (1973a).

The European context

The European Union (EU), now comprising twenty-seven countries, was established in 1992 by the Treaty on European Union (the so-called Maastricht Treaty). The common language to describe beef carcasses, the language of carcass classification, was already in existence when the EU was formed and the scheme was formulated in 1981 under the auspices of the European Economic Community (later, simply the European Community), one of the three founding 'pillars' of the EU. So the beef carcass classification scheme used throughout the EU is the EC one and is widely known as 'the EUROP system.

Developments occurring prior to the adoption of the EUROP system

Although the origin of the 'classification movement' in Europe cannot, perhaps, be identified easily and with certainty, several countries that later became members of the EU had developed, and were executing, their own systems in the late 1960s/early 1970s; there were schemes in operation in Germany, France, Ireland and Britain. By the late 1970s there were also Dutch and Danish systems in use. Interestingly, these early schemes were all based on the same carcass characteristics: weight, age, gender, fatness and conformation. The number of classes was not the same between countries: Britain had five classes for fatness as did France for older cattle, whilst Germany had three and Ireland were proposing seven. For conformation, Britain had five classes, France six and Germany four. Ireland did differ in their approach to measuring conformation, preferring ten classes based on the objectively determined weight to length ratio. A more comprehensive review of the situation in various countries is in Kempster *et al.* (1982).

Experience in using a scheme often leads to revision. For example, in Britain the original scheme, as noted above, comprised five fatness and five conformation classes. However, it soon became apparent that too high a proportion of carcasses were falling in the modal class 3 and so it was subdivided into 3L (Low) and 3H (High). There was also a need to accommodate, separately, carcasses of extremely poor conformation, so the Meat and Livestock Commission introduced a 'Z' class (Table 3).

Table 3. The modified classification grid used in Britain pre-1981 (MLC) and its alignment with the European Scheme (EEC).

	Fatness								
	MLC		1		2	3	3H	4	5
		EEC [a]	1	2	3	4L	4H	5L	5H
Conformation	5	E							
		U+							
	4	U							
	3	R							
	2	O							
	1	O-							
	Z	P							

[a] These class subdivisions were the ones used in Britain in 1981. The conformation subclasses at the present time are E, U+, -U, R, O+, -O, P+, -P.

Although the methods used to classify beef carcasses in these countries were similar, bigger differences prevailed in the method of grouping cattle according to age and gender. For example, in the UK there was differentiation between bulls, steers (castrated males), heifers (unbred young females) and cows, with an optional distinction between young stock (based on dentition - up to two permanent incisors) and the rest; in Germany there were nine categories (calves, young cattle, steers, heifers, young bulls, bulls, very young cows, young cows and cows), age identification being based on the visual assessment of ossification of certain cartilages and on the colour of bones. Harrington (1973b) considered that the greatest obstacle to the formation of a common European system was this variation in categorising carcasses according to age and gender, rather than the method of describing fat and conformation. Another hurdle was the definition of the carcass and its weight recording, even within national schemes. All of the schemes mentioned above included cold weight but this was usually estimated from the hot side weight recorded at the end of the slaughter line and a rebate applied, either a set percentage deduction or a sliding scale. However, in the French scheme 'hot' weight was recorded up to three hours after slaughter and not necessarily within one hour, as was the norm. Variation in dressing practice was even more important and clearly distorted the market, making it unfair to farmers who were paid partly on the basis of weight and to traders buying and selling a non-standard product.

In Sweden, a revised beef carcass classification scheme was introduced in 1969 and all carcasses were assessed from that time. The system used was based on a score card: there were two classes - a fatness class and a 'basic' class, the latter being partly dependent on fleshing (visually assessed out of a possible 13 points) and partly on lean meat quality (4 points). The sum of these scores allocated carcasses to one of seven basic classes whist the fatness scale was of five classes of fat thickness ranges, measured objectively. This approach was later changed to one in which fleshing and fatness were classified in six categories (Kempster *et al.*, 1982)

The very first attempt to achieve some sort of standardisation and harmonization in carcass assessment occurred about ten years after the formation of the original EEC in 1958 when Belgium, France, Italy, Luxembourg, the Netherlands and West Germany joined under the Treaty of Rome to form the 'Common Market'. Experts from these six countries agreed a standard description of beef carcasses, partly based on the intact side (half carcass) and partly on observations made after cutting (Kempster *et al.*,1982). However, it was not until the 1970s that debate on a common EEC scheme was discussed in earnest. Despite the inconsistencies in carcass dressing and weighing, described above, Harrington (1973b) concluded from his comparative study that 'this … suggests that a common European beef carcase classification scheme for price reporting purposes is quite feasible, although it would probably have to be quite complicated in terms of the subdivisions and class combinations to embrace the whole spectrum of European beef quality'. He suggested that the time was right for meat trade associations throughout Europe to liaise with their technical bodies (like the Meat & Livestock Commission in Britain), begin the process of drawing up a common system and not wait for Commission bureaucrats in Brussels to decide that a common scheme was necessary for deadweight price reporting. However, at a seminar held in Abano Terme, Italy, some five years later, under the auspices of the Commission of the European Communities, the same author reported the recommendations of one of three working groups and stated that 'the difficult task of formulating a common beef carcass classification system should be begun without further delay. This should not be a charge on the research budget as the prime purpose is to achieve progress on the accurate reporting of deadweight prices which is essential to efficiency in the market' (Harrington, 1978a). So it is clear that the rate of progress was slow, mainly because of national protectionism and political posturing, a fact acknowledged by H. de Boer, the well-known researcher from the Research Institute for Animal Husbandry in Zeist, The Netherlands (who, possibly more than anyone else, strove for harmonisation in European carcass evaluation) in his foreword to the book by Kempster *et al.* (1982), in which he states 'Many years of personal involvement in coordination at the European level have shown the consequent entanglements and laborious efforts required to achieve harmonisation'.

The European Commission had recognized, by 1978, that price reporting on a deadweight basis, i.e. carcass-based, as opposed to a live cattle system, was becoming increasingly necessary and to have comparability of prices, a uniform classification scheme was essential. This would enable more precise identification of those carcasses to be taken off the market in times of surplus, i.e. intervention. But there were many hurdles to overcome and these were set out by Harrington (1978b). He identified as the biggest problem the potential confusion of objectives: does the best system for pan-European price reporting also serve best the needs of the individual farmer in determining choice of breeds or the local butcher in satisfying his customers' requirements? Another related problem was that a lot of effort had already been put in to develop national schemes and there would likely be resistance to an EEC scheme that would be workable, but only at the expense of the internal value of the national scheme.

The role of research and the EAAP reference method

In the 1970s there were two publications by Dumont and his co-researchers in INRA, France, on the assessment of conformation (Dumont *et al.*, 1975) and fatness (Roy and Dumont, 1975). These gave very detailed descriptions of carcass shape and fat deposition, and indicated the need to assess shape from a number of orientations of the carcass with respect to the assessor, and to assess fatness in a number of carcass regions. Although these were too detailed for commercial application, they did embody some important principles that had been accepted by the Study Commission on Cattle Production of the EAAP and incorporated in the EAAP Standard Method of beef carcass assessment (de Boer *et al.*, 1974). The intended use of this manual was to help researchers to describe carcasses used in experiments in a uniform way, so that results could be interpreted clearly and comparisons made across studies. It included a photographic reference of carcasses depicting fatness and conformation and covered the range of carcass types found throughout Europe. There were five main classes for both fatness and conformation, with each main class subdivided into three (Table 4). Harrington (1978b) proposed that it would be sensible to use the EAAP scheme as a basis for discussion of classification in an EEC context and he suggested that such discussions would centre on the number of subdivisions of the range to make the scheme commercially viable. This is, in fact, what happened and the basic structure of five main classes for both fatness and conformation, with suitable subdivisions as appropriate, was seen as being adequate to describe the very variable European beef population and it now forms the backbone of the EUROP scheme. The relevance of carcass profiles, proposed by Houdinière (cited by Dumont *et al.*, 1975) and incorporated in their method) was also appreciated and incorporated in the EUROP scheme, as was the importance of thoracic fat development (Roy and Dumont, 1975). The significance of the EAAP and underpinning studies was, therefore, considerable.

Table 4. The five main classes and fifteen subclasses for assessment of fleshiness and fatness in the EAAP Reference Methods for the assessment of carcass characteristics in cattle.

1			2			3			4			5		
1-	1	1+	2-	2	2+	3-	3	3+	4-	4	4+	5-	5	5+

The Community scale for the classification of carcasses of adult bovine animals.

The purpose of introducing a common classification, as far as the European Commission was concerned, was to facilitate intra-community trade and to provide a basis for support measures. It provides an index of the saleable meat yield of a carcass and also provides some insight into factors that affect the appearance of the meat at retail. The EEC Beef Carcass Classification Scheme's basis is in Council Regulation 1208/81 and it has subsequently undergone several revisions and amendments with the latest being Commission Regulation (EC) No 103/2006, the self-declared aim of which was to ensure the uniform classification of the carcases of adult bovine animals in the Community by making the definitions of classes of conformation and fat cover more precise.

The description of the main classes is shown in Table 5 and 6. Within each class there may be up to fifteen subclasses (cf. Table 4), and each member country can decide how many subclasses are necessary to adequately describe its national population. The main classes are represented in photographic standards, for both conformation and fatness. In describing carcasses, the conformation class is given first, i.e. a letter followed by a number (e.g. R3).There are statutory checks on the

Table 5. The Community scale for the classification of carcasses of adult bovine animals: classes for conformation with their corresponding key carcass descriptions.

Conformation		
Class	Essential part	Description
S (superior) All profiles extremely convex; exceptional muscle development (double-muscled carcase type)	Round	very highly rounded; doublemuscled; visibly separated seams; topside spreads very markedly over the symphysis pelvis
	Back	very wide and very thick, up to the shoulder; rump very rounded
	Shoulder	very rounded
E (excellent) All profiles convex to super-convex; exceptional muscle development	Round	very rounded; topside spreads markedly over the symphysis pelvis
	Back	wide and very thick, up to the shoulder; rump very rounded
	Shoulder	very rounded
U (very good) Profiles on the whole convex; very good muscle development	Round	rounded; topside spreads over the symphysis pelvis
	Back	wide and thick, up to the shoulder; rump rounded
	Shoulder	Rounded
R (good) Profiles on the whole straight; good muscle development	Round	well-developed; topside slightly rounded
	Back	still thick but less wide at the shoulder; rump slightly rounded
	Shoulder	fairly well-developed
O (fair) Profiles straight to concave; average muscle development	Round	average development to lacking development
	Back	average thickness to lacking thickness; rump has straight profile
	Shoulder	average development to almost flat
P (poor) All profiles concave to very concave; poor muscle development	Round	poorly developed
	Back	narrow with bones visible
	Shoulder	flat with bones visible

Table 6. The Community scale for the classification of carcasses of adult bovine animals: classes for fatness with their corresponding key carcass descriptions.

Fatness		
Class	Cover (subcutaneous) fat	Thoracic fat
1 (low)	None up to low fat cover	no fat
2 (slight)	Slight fat cover, flesh visible almost everywhere	the muscle is clearly visible between the ribs
3 (average)	Flesh, with the exception of the round and shoulder, almost everywhere covered with fat	slight deposits of fat but the muscle is still visible between the ribs
4 (high)	Flesh covered with fat but still partly visible on the round and shoulder; the seams of fat on the round are prominent.	distinctive fat deposits; the muscle between the ribs may be infiltrated with fat
5 (very high)	Entire carcass covered in fat; the round is almost completely covered with fat, so that the seams of fat are no longer clearly visible.	heavy fat deposits; the muscle between the ribs is infiltrated with fat

performance of assessors: on-the-spot inspections are carried out by a Community inspection committee composed of experts from the Commission and experts appointed by the Member State. This Committee reports back to the Commission on checks carried out.

Recent advances

Although strictly outside the remit of this paper, mention should be made of important recent developments in beef carcass classification that give some insights of the future. The advantages of moving to an objective system for classifying carcasses have long been recognised but no method for beef has performed to a standard deemed adequate by the trade or regulatory bodies. Quite recently, however, video image analysis (VIA) techniques have been developed and are being used to classify beef (and sheep) carcasses. Put simply, they mimic the human assessor by analysing an image of the carcass. Particularly noteworthy has been the work in Ireland which, in pursuit of the introduction of automated classification, carried out two substantial trials of three VIA systems in 1999 and 2000. In July 2003 the regulation was changed to allow mechanical grading systems to be used provided they were sufficiently accurate (Commission of the European Communities, 2003: EC 1215/2003). This contains the rules for carrying out authorisation trials and the statistical criteria that must be met. All three machines tested met the requirements and have now been licensed for use by the Irish beef slaughtering industry. In 2004 Ireland became the first country to have VIA systems authorised and 24 systems are now in operation (Allen, 2005). Uptake in other countries is to be expected.

Acknowledgements

The author is particularly grateful to Neil Hudson, Librarian at the Meat & Livestock Commission, for the acquisition of some of the papers cited in this review.

References

Allen, P., 2005. Evaluating video image analysis (VIA) systems for beef carcass classification. The Science of Beef Quality, Eighth Annual Langford Food Industry Conference, British Society of Animal Science, 9-12.

Commission of the European Communities, 1984. Commission Regulation (EEC) No 3220/84.

Commission of the European Communities, 2003. Commission Regulation (EC) No. 1215/2003 (OJ No. L169, 32).

Council of the European Union, 2006. Council Regulation (EC) No 1183/2006.

De Boer, H., B.L. Dumont, R.W. Pomeroy and J.H. Weniger, 1974. Manual on E.A.A.P. reference methods for the assessment of carcase characteristics of cattle. Livestock Production Science, Volume 1, 151-164.

Dumont, B.L., P. le Guelte and J. Sornay, 1975. Le jugement de la conformation des carcasses de bovines. INRA/ITEB, 26pp.

Harrington, G., 1973a. Some technical problems in developing a beef carcase classification system. Institute of Meat Bulletin, Volume 80, 21-26.

Harrington, G., 1973b. Classification developments in Britain and Europe. Proceedings of the Meat and Livestock Commission National Meat Conference, Stratford, England. Mimeograph, 25-36.

Harrington, G., 1978a. Report of third interdisciplinary group discussion. In J.C. Bowman and P. Susmel (eds.) The Future of Beef Production in the European Community, Martinus Nijhoff, The Hague, The Netherlands, 505-509.

Harrington, G., 1978b. Problems in devising a European beef carcase classification system. European Congress for Improved Beef Productivity. Elanco Products Ltd. Paris, France.

JMGA, 1988. New beef carcass grading standards. Japan Meat Grading Association, Tokyo, Japan.

Kempster, A.J., A. Cuthbertson and G. Harrington, 1982. Carcase Evaluation in Livestock Breeding, Production and Marketing. Granada, London, UK.

Meat & Livestock Commission, 1975. Progress on beef carcase classification. Marketing and Meat Trade Technical Bulletin no. 22. MLC, Bletchley, UK.

Meat Standards Australia, undated. http://www.mla.com.au/NR/rdonlyres/68780D90-BD35-48D2-921C-BBF7BA49C4B7/0/MSAbeef4ppfinal.pdf

Pomeroy, R.W., 1977. Carcass classification. Meat Research Institute Biennial Report 1975-1977. Agricultural Research Council, Bristol, UK, 35-41.

Price, M.A., 1995. Development of carcass grading and classification systems. In Morgan Jones, S.D. (ed.), Quality and Grading of Carcasses of Meat Animals, CRC Press, Boca Raton, FL, USA, 173-199.

Roy, G. and B.L. Dumont, 1975. Methode de jugement descriptif de l'etat d'engraissement des carcasses de bovines adultes. Revue de Médicine Véterinaire, Volume 126, 387-400.

Starke, J.S. and D.M. Joubert, 1961. A score card for lamb and mutton carcasses. Journal of Agricultural Science Cambridge, Volume 57, 319.

United States Department of Agriculture, 1997. United States Standards for Grades of Carcass Beef. 17pp.

Williams, D.R., 1969. The visual description of carcasses. Journal of Agricultural Science Cambridge, Volume 73, 495-499.

Williams, D.R., 1972. Visual assessment of beef and lamb. World Review of Animal Production, Volume 8, 87-96.

Evaluation of sheep carcass quality

Francesca Maria Sarti and F. Panella
Università degli Studi di Perugina, Dipartimento di Biologia Vegetale e Biotecnologie Agroambientali e Zootecniche, Via Borgo XX Giugno 74, 06121 Perugia, Italy

Abstract

The quality is difficult to define because it is a mix of sanitary, technological and organoleptic components. In Europe there are different kinds of products: in the Central and Northern part lambs are produced, on the contrary in the Mediterranean regions heavy, light and milk lambs are more popular; the first one's are classified according to the SEUROP grid and the other by the A grid. The countries where some special carcass classification methods are developed are: Sweden, Netherlands, UK, and Scotland. Anyway in France, Sweden, Finland and German the SEUROP classification it's a statutory requirement. Around the world a big effort on this topic has been made by the big producers such as Australia and New Zealand in order to find a grading system able to define the price according to the carcass quality.

Keywords: sheep, carcass evaluation, in vivo, ex vivo

Introduction

It is difficult to define quality because it is a mix of sanitary, nutritional, technological and organoleptic components. Moreover, the meaning of quality of carcasses changes in several world areas according to the local customs. As a matter of fact, the fat content preferred by the consumers is different according to geographical areas, culinary uses and social class; in fact in the less developed countries, or where the grilled meat is very popular, fat is liked (Gall, 1982). In many countries (Africa, South America) mature animals are preferred. This tradition and liking come from the need to take advantage of the sheep whole growth potential.

The blackhead sheep, which is found in many African countries (Morocco, Algeria, Tunisia, Libya, Ethiopia, Kenya, Somalia), is slaughtered when it is around a year old (30-35 kg) and gives a low conformed, but very fat carcass. This product is also appreciated in Arabian countries. Also in South American countries (Peru, Argentina, Chile) lambs are slaughtered at nearly 30-50 kg of live weight (Rubino *et al.*, 1999).

Moreover, this trait can be assessed on either carcasses or live animals for slaughter and breeding; nevertheless, there is no single universally accepted definition of the term carcass conformation across the sheep industry.

Sheep carcass evaluation in Europe

In Europe there are different kinds of sheep products in relation to location; in the central and northern parts the main product is the heavy lamb (over 13 kg carcass) called 'mutton' (UK), 'mouton' (France), 'ovino mayor' (Spain) and 'agnello pesante' (Italy). In the same countries, lamb (7-13 kg carcass) called 'agneau' (France), 'cordero' (Spain) and 'agnello leggero' (Italy) is also produced. In the Mediterranean region (Spain, Italy, Greece, …), because of the use of ewes' milk, the milk lamb (up to 7 kg carcass) is the favourite; this product is called 'agneau de lait' (France), 'agnelet' (Basque Countries), 'lechal' (Spain) and 'agnello da latte' (Italy).

This variety in European production makes necessary to evaluate different kinds of carcasses (EC Reg. 3013/89) according to weight. The SEUROP grid is used for heavy lambs; there are six classes of conformation (decreasing from S to P) and five classes of fat (increasing from 1 to 5). The other carcasses in some countries are classified by 'A or Mediterranean grid' (EC Reg. 2137/92) according to meat colour (three classes) and fat quantity (four classes).

Sweden had one of the earliest classification systems in place. By 1964, 70-80% of the meat sold in the Stockholm meat market was sold by description without inspection. All carcasses produced in Sweden intended for sale on the open market have to be classified, according to rules issued by the Swedish Board of Agriculture; carcasses are classified according to age (suckler lamb, lamb or sheep), conformation and fatness into seven classes from M (completely lacking) through to L (light), N (normal), B (3-5 mm), C (5-10 mm), D (10-15 mm) and E (more than 15 mm). For the fatter classes a linear measurement of fat depth over the loin muscle is made between the 10th and 11th rib by making a cut and inserting a rule. This measure is similar to C site fat measurement rather than Grade Rule (GR) tissue depth used in the Australian fat scoring system.

A carcass grading system is available in the Netherlands but in practice it is hardly used. The sheep industry is very small (less than 700,000 sheep, lambs and goats are slaughtered in the Netherlands annually). Most of the domestic consumption of lamb is by people originating from Morocco and Turkey and living in the Netherlands who prefer a light weight carcass. A small amount of lamb also goes into the food service industry. Most trading is done via stock traders to the slaughter houses making it very difficult to inform farmers of SEUROP results, consequently they are not willing to pay for it.

The UK first implemented a classification system in 1973, this used 5 numeric fat classes based on visual assessment and four conformation scores, 'Z' very poor, 'C' poor, average (no letter code) and 'E' extra. More recently the UK has adopted the EUROP classification for conformation (there is no 'S' grade used for sheep in the UK) and retained the numeric (1-5) assessment for fatness. The scheme is designed to describe the main characteristics of the carcass without attributing any qualitative judgement. The main objective is to indicate the likely yield of the carcass; of course the actual yield will depend on the butchery specifications in place in a particular business.

The selection of the particular approach (EUROP and numeric fat scale) was based on the assumption that there would ultimately be a Europe-wide classification scheme for sheep. The classification scheme was therefore based on the existing Europe-wide scheme for cattle as it was felt appropriate to devise a scheme that was likely to match any future European requirements. The reason for the introduction of the EUROP scheme was the desire for a common language for describing the carcass to improve industry communication. The main argument was that farmers were not being paid fairly for better animals and therefore there was little incentive for breed improvement. Over fatness has always been considered as the main issue with carcass quality in the UK. Introduction was not driven by any concerns over consumer satisfaction and there was no intention for it to be used as a prediction of eating quality.

Lamb classification in the EU is on a voluntary basis and is not a requirement under EU law, unlike beef where it is a statutory requirement. French, Sweden, Finland and German governments have adopted the SEUROP grid as obligatory; while Denmark, Finland and UK utilize this grid occasionally. The objective of this classification was to ensure greater transparency in transactions and to encourage production of carcasses of better quality (less fat and greater yield) (Wiese and Wood, 2006).

In Europe a new research in this field is carried out by the Scottish Agricultural College (SAC). SAC is to lead a new project, which will use modern research technology to validate genetic factors that could lead to improvements in the lean meat yield from sheep. The research will also further test and help to calibrate a new video image scanning and analysis (VISA) system for the objective evaluation of lamb carcasses. This project is the result of a major government and industry funded collaboration which will see researchers from across the UK working with representatives of the lamb breeding and meat industry to evaluate new DNA based genetic test methods and automatic carcass evaluation systems, that can be built into selection programmes for sheep breeding.

The project will have the short acronym 'Sheep QTL and VISA' and is jointly funded by BBSRC (Biotechnology and Biological Sciences Research Council), Defra, EBLEX (English Beef & Lamb Executive), QMS (Quality Meat Scotland), HCC (Hybu Cig Cymru/Meat Promotion Wales), Livestock and Meat Commission for Northern Ireland, Scottish Association of Meat Wholesalers and Catapult Systems Ltd through the Sustainable Livestock Production LINK programme. The four-year project involves comprehensively evaluating the effects of three specific 'muscling' segments of chromosomes which contain naturally occurring mutations that affects muscle growth (QTLs = quality trait loci) and which can be determined from a DNA blood test. The researchers will use Catapult's DNA testing services to classify animals according to the number of copies of QTL known to affect muscle growth in Texel and Dorset sheep which have been identified from previous research in both the UK and New Zealand.

Computer tomography and VISA along with other methods will be used to evaluate the relationships between QTL presence and carcass and meat quality, with particular reference to lean content, muscle volume and shape. This new technology promises to provide a transparent value-based marketing system with the potential to send clear and accurate market signals from the consumer upward through the whole supply chain to primary producers. Such a system must have the means to identify the value of individual carcasses based on carcass conformation and composition, yield of lean meat and individual cuts. Given their potential benefits to the supply chain, such systems need detailed evaluation, and in particular they need to be calibrated for the whole sheep sector within the UK (SAC, 2006).

Sheep carcass evaluation around the world

Around the world there are various methods for predicting lamb carcass composition *in vivo* and *ex vivo* as reported in a review of Stanford (Stanford *et al.*, 1998). The keys to changing carcass composition to better meet consumer demand are methods of evaluating body composition *in vivo*.

Visual appraisal in combination with condition scoring (manual assessment of fatness) is the most rapid and inexpensive method for prediction of body composition *in vivo* (Kempster, 1984). However, the large variation between breeds in the proportion of fat stored subcutaneously (Fahmy *et al.*, 1992) limits the usefulness of this method.

In vivo techniques commonly use live weight as the standard to which other predictors of body composition are compared (Kempster, 1984; Simm, 1992) although live weight may be difficult to accurately measure due to the influence of gut fill and fleece length/hydration. The usefulness of live weight as a predictor of body composition is limited by difficulties in assessing an animal's stage of maturity which can be influenced by genotype, nutrition, disease, physical environment, level of activity, social environment and age (Taylor, 1965).

Prior to development of technologies enabling *in vivo* prediction of carcass composition, a number of *linear measurements* (shoulder height, heart girth, body length, etc.) were evaluated as predictor of body composition in sheep (Orme *et al.*, 1962; Orme, 1963; Cunningham *et al.*, 1967), but were found to be of marginal utility in lambs of varying age, sex or breed type.

Ultrasound equipment converts electrical pulses to high-frequency sound waves which were reflected from the boundaries between tissues of different bioacoustic densities (Houghton and Turlington, 1992). Two types of ultrasound equipment are used: (a) mode machines, available since the 1950's, which measure echo amplitude against time, with the distance between echoes being related to the distance between successive tissue interfaces (Simm, 1983); (b) model or real-time machines developed in the early 1980's, where 'grey scales' measure echo intensity in a two-dimensional scan (Stouffer, 1988). The velocity of ultrasound through soft tissues is also used to predict body composition (Miles *et al.*, 1991), offering the advantage of absolute values instead of images requiring subjective interpretation. The limited utility of ultrasound was attributed to the small size and lack of variation in subcutaneous fat thickness and *longissimus* muscle area in sheep as compared to cattle and pigs (Houghton and Turlington, 1992).

The equipment used in *X-ray computed tomography* (CT) was first developed for human medicine (Vangen, 1989). An X-ray generator and X-ray detectors are rotated around the subject, firing pulses of radiation and measuring the amount of radiation transmitted trough the subject (Simm, 1992). The rate of attenuation of the X-rays allows computerized calculation of densities in a cross section of the subject, the densities being standard values which vary from -1000 for air to +1000 for bone (Standal, 1984). *In vivo* use of CT is applicable only for smaller livestock such as sheep, goats, chickens and pigs due to the human-scale of the equipment (Vangen, 1989).

A *nuclear magnetic resonance* (NMR) machine consists of an electromagnet with a central opening large enough for a human. The strong magnetic field tends to induce resonance of protons in the subject (Wells, 1984). Contrary to CT, there are no standardized values in NMR due to changes in conditions and parameters between measurements (Groeneveld *et al.*, 1989). However, NMR has additional capabilities compared to CT including evaluation of muscle metabolism and prediction of carcass quality attributes such as water holding capacity (Monin and Renou, 1989). The accuracy of NMR in predicting body composition is thought to be superior to that of CT (Groeneveld *et al.*, 1984; Simm, 1992), although CT and NMR were found to be of equal value in determining adipose tissue volumes of rats (Ross *et al.*, 1991).

Other techniques for prediction of body composition, presently used in human medicine, include *dual-photon absorptiometry* (Mazess *et al.*, 1990), *dual X-ray absorptiometry* (Dalsky *et al.*, 1990) and *underwater weighing* (Wang *et al.*, 1989).

Carcass composition assessment *ex vivo* serves three functions: (l) assigns carcass value; (2) allows sorting of carcasses for further processing or fresh meat merchandising and (3) transfers information back to the production sector, hopefully ensuring that carcasses meet consumer demand. In the more developed areas of the world carcass quality is evaluated by *grids* that give the price of the product. The quality indicated on many labels is made according to some grid evaluation. In the USA (Boggs and Merkel, 1981), carcasses must be identified as to their sex or class and their maturity group or kind. The classes or sex group of sheep are ewe (female), wether (castrated male) and ram (male). The classes are then further subdivided by kind or maturity groups into lambs (up to 14 months of age), yearlings (12-15 months) and mutton (older than 24 months). Each of the kinds or maturity groups contains ewes, wethers and rams.

Lamb carcasses may be quality graded, yield graded or both. Quality grade is an estimation of eating characteristics, i.e. tenderness, juiciness and flavour, while yield grade is an estimation of boneless, closely trimmed, retail cuts form the leg loin, rack and shoulder. In New Zealand the current measurement used to define export lamb classes and carcass composition (meat and fat) is the Grade Rule (GR) (Kirton *et al.*, 1992). It places lamb carcasses into one of 14 classification boxes based on weight and fatness classification. Eight of the boxes (A, YL, YM, YX, PL, PM, PX, PH) contain carcasses that can be exported; the remaining boxes (TL, TM, TH, FL, FM, FH) contain carcasses that should be cut and trimmed before export. The present subjective system also sorts carcasses into five fatness bands (A, Y, P, T, F) (Rubino *et al.*, 1999).

The utility of subjective methods for evaluating carcass composition has been largely dependent of the population of lambs evaluated. Fat has a lower density than other carcass components and the determination of *carcass specific gravity* [weight in air/(weight in air-weight underwater)] was the subject of early investigations (Kirton and Barton, 1958; Timon and Bichard, 1965). In these studies, carcass specific gravity was found to be equal to dressing percentage (hot carcass weight (HCW)/live weight) as a predictor of carcass fat content, although specific gravity was not deemed sufficiently accurate for individual carcass determination due to a high RSD (2.98 to 3.2 for % carcass fat) and its reduced accuracy at lower levels of fatness. Comparing carcass weight, *specific gravity* and *dressing percentage*, Barton and Kirton (1958) found carcass weight to be the superior predictor of carcass fat content in sheep as it was not subject to as many errors in measurement as was specific gravity and was not influenced by variations in gut fill as was dressing percentage.

Optical probes objectively measure fat and muscle depths and are routinely used to measure GR according to New Zealand export lamb grading regulation (Price, 1995). Optical probes consist of a light-emitting diode which illuminates the meat from under an optical window. Detectors respond to an increase in reflected light when the optical window passes from muscle to fat as the probe is withdrawn from the carcass (Swatland, 1995). As reported by Kirton *et al.* (1995), probes currently available for prediction of lamb carcass composition include the Hennessy Grading Probe (Hennessy Grading System, Auckland, NZ), the AUS-Meat Sheep Probe (SASTEK, Hamilton, Queensland, Australia), the Swedish FTC lamb probe (FTC Sweden, Upplands, Väsby, Sweden) and the Ruakura GR Lamb Probe (Hamilton, NZ). Only the AUS-Meat probe is capable of functioning at chain speeds of nine to 10 carcasses per minute (Cabassi, 1990; Hopkins *et al.*, 1995).

Wood *et al.* (1991) described *Video Image Analysis* (VIA) as a system capable of objectively measuring carcass conformation, with equipment including a video camera, controlled lighting of the carcass and computer/software necessary to digitize the video image; VIA shows potential as an objective, accurate, yet cost-effective method of evaluation of lamb carcass composition.

Lean tissue is approximately 20 times more conductive of electricity than fat or bone because of higher concentrations of water and electrolytes (Funk, 1991). Based on this principle, carcasses passed through an electromagnetic coil generate a relative energy absorption curve. Areas under parts of the curve and differences between positions on the curve are therefore related to lean mass (Swatland, 1995). Although this technology has been used on live pigs (EMME electronic meat measuring equipment, EMME, Phoenix, AZ), movement of the pigs could led to inaccurate estimates of lean content (Forrest *et al.*, 1991). Using TOBEC (Meat Quality, Springfield IL), electrical conductivity measurements and carcass length were able to predict carcass lean percentage in lambs with a reasonable degree of accuracy (R^2=0.78, RSD=1.71%), although carcass-position within the scanner, carcass temperature and geometric orientation of the carcass were recognized as sources of error (Berg *et al.*, 1994).

Another method dependant on transmission of electric current through a carcass, *bioelectrical impedance* (BIA) measures are related to conductor, cross-sectional area and signal frequency; leading to the hypothesis that a fat lamb should impede the transmission of electrical current to a larger extent than a lean lamb (Berg *et al.*, 1996). Two pairs of transmitter detector electrodes (21 gauge needles) are located in an anterior to posterior sequence along the full length of the animal's back (Swatland, 1995). Impedance measurements include resistance and reactance which are calculated by transmitting alternating current between the outer two electrodes and measuring the voltage drop between the inner two detector electrodes (Berg and Marchello, 1994). An advantage to bioelectrical impedance is that measurements can be made in live animals as well as carcasses (Berg and Marchello, 1994; Berg *et al.*, 1996), although the invasiveness of the procedure as well as its low precision would not favourably compare to other relatively inexpensive *in vivo* methods such as ultrasound.

References

Barton, R.A. and A.H. Kirton, 1958. Carcass weight as an index of carcass components with particular reference to fat. Journal of Agricultural Science, Volume 50, 331-334.

Berg, E.P. and M.J. Marchello, 1994. Bioelectrical impedance analysis for the prediction of fat-free mass in lambs and lamb carcasses. Journal of Animal Science, Volume 72, 322-329.

Berg, E.P., J.C. Forrest, D.L. Thomas, N. Nusbaum and R.G. Kauffman, 1994. Electromagnetic scanning to predict lamb carcass composition. Journal of Animal Science, Volume 72, 1728-1736.

Berg, E.P., M.K. Neary, J.C. Forrest, D.L. Thomas and R.G. Kauffinan, 1996. Assessment of lamb carcass composition from live animal measurement of bioelectrical impedance or ultrasonic tissue depths. Journal of Animal Science, Volume 74, 2672-2678.

Boggs, D.L. and Merkel, R.A., 1981. Live animal: carcass evaluation and selection manual. Iowa: Michigan State University. 199p.

Cabassi, P., 1990. The prediction of lamb carcass composition from objective measurements of fatness taken at slaughter chain speed with the Aus-Meat sheep probe. Proceedings of Australian Society of Animal Production, Volume 18, 164-167.

Cunningham, N.L., Z.L. Carpenter, G.T. King, O.D. Butler and J.M. Shelton, 1967. Relationship of linear measurements and certain carcass characteristics to retail value, quality and tenderness of ewe, wether, and ram lambs. Journal of Animal Science, Volume 26, 683-687.

Dalsky, G.P., W. Kraemer, A.E. Zetterlund, B. Conroy, A. Fry, J.O. Judge and J. Smith, 1990. A comparison of methods to assess body composition. Proceeding American College of Sports Medicine, Salt Lake City, UT, USA, May 22-25.

Fahmy, M.H., J.M. Boucher, L.M. Poste, R. Gregoire and J.E. Comeau, 1992. Feed efficiency, carcass characteristics and sensory quality of lambs with or without prolific ancestry fed diets with different protein supplements. Journal of Animal Science, Volume 70, 1365-1374.

Forrest, J.C., C.H. Kuei, W. Chen, R.S. Lin, A.P. Schinckel, P. Walstra, H. Kooper and M.D. Judge, 1991. Electromagnetic scanning: carcass evaluation. Proceeding Symposium on Electronic Evaluation of Meat in Support of Value-Based Marketing, March 27-28, Purdue University, West Lafayette, IN, USA, 85-112.

Funk, R., 1991. Electromagnetic scanning: basis and recent advances in technology. Proceeding Symposium on Electronic Evaluation of Meat in Support of Value-Based Marketing, March 27-28, Purdue University, West Lafayette, IN, USA, 73-84.

Gall, C.F., 1982. Goat carcass composition. Proceeding 3rd International Conference on Goat Production and Disease, 10-15 January, Tucson, USA, 472-487.

Groeneveld, E., M. Henning and E. Kallweit, 1989. Growth patterns and carcass evaluation in pigs NMR measurements. In: Kallweit, E., M. Henning and E. Groeneveld (Eds.), Application of NMR Techniques on the Body Composition of Live Animals, Elsevier, New York, USA, 137-148.

Hopkins, D.L., M.A. Anderson, J.E. Morgan and D.G. Hall, 1995. A probe to measure GR in lamb carcasses at chain speed. Meat Science, Volume 39, 159-165.

Houghton, P.L. and L.M. Turlington, 1992. Application of ultrasound for feeding and finishing animals: a review. Journal of Animal Science, Volume 70, 930-941.

Kempster, A.J., 1984. Cost-benefit analysis of in vivo estimates of body composition in meat animals. In: Lister, D. (Ed.), In Vivo Measurement of Body Composition in Meat Animals, Elsevier, New York, USA, 191-203.

Kirton, A.H. and R.A. Barton, 1958. Specific gravity as an index of the fat content of mutton carcasses and various joints. New Zealand Journal of Agricultural Research, Volume 1, 633-641.

Kirton, A.H., G.J.K. Mercer, D.M. Duganzich and A.E. Uljee, 1995. Use of electronic probes for classifying lamb carcasses. Meat Science, Volume 39, 167-176.

Kirton, A.H., Mercer, G.J.K. and Duganzich, D.M., 1992. A comparison between subjective and objective (carcass weight plus GR or the Hennessy Grading Probe) methods for classifying lamb carcasses. Proceedings of the New Zealand Society of Animal Production, Volume 52, 41-44.

Mazess, R.B., H.S. Barden and E.R. Olrich, 1990. Skeletal and body composition effects of anorexia nervosa. American Journal of Clinic Nutrition, Volume 52, 438-441.

Miles, C.A., G.A.J. Fursey, A.V. Fisher and S.J. Page, 1991. Estimation of lamb carcass composition from measurements of the speed of ultrasound in the soft tissues of live animals and carcasses. Meat Science, Volume 30, 245-256.

Monin, G. and J.P. Renou, 1989. Spectroscopy and meat quality. In: Kallweit, E., M. Henning and E. Groeneveld (Eds.), Application of NMR Techniques on the Body Composition of Live Animals, Elsevier, New York, USA, 121-133.

Orme, L.E., 1963. Estimating composition from linear measurements, live probe and body weight. Annals of the New York Academy of Science, Volume 110, 307-310.

Orme, L.E., R.E. Christian and T.D. Bell, 1962. Live animal and carcass indices for estimating the carcass composition in lambs. Journal of Animal Science, Volume 21, 666.

Price, M.A., 1995. Development of carcass grading and classification systems. In: Morgan Jones, S.D. (Ed.), Quality and Grading of Carcasses of Meat Animals, CRC Press, Boca Raton, FL, USA, 173-199.

Ross, R., L. Leger, R. Guardo, J. de Guise, B.G. Pike and J. de Guise, 1991. Adipose tissue volume measured by magnetic resonance imaging and computerized tomography in rats. Journal of Applied Physiology, Volume 70, 2164-2172.

Rubino, R., P., Morand-Fehr, C. Renieri, C. Peraza, F.M. Sarti, 1999. Typical products of the small ruminant sector and the factors affecting their quality. Small Ruminant Research, Volume 34, 289-302.

SAC Scottish Agricultural College, 2006. Sheep Breeding Research to Link New Genetic Factors and Carcass Quality. www.sac.ac.uk/news/newsarchives/carcassquality.

Simm, G., 1983. The use of ultrasound to predict the carcass composition of live cattle – a review. Animal Breeding Abstract, Volume 51, 853-875.

Simm, G., 1992. Selection for lean meat production in sheep, In: Speedy, A.W. (Ed.), Progress in Sheep and Goat Research CAB International, Wallingford, UK, 193-215.

Standal, N., 1984. Establishment of a CT facility for farm animals. In: Lister, D. (Ed.), In Vivo Measurement of Body Composition in Meat Animals, Elsevier, New York, USA, 43-51.

Stanford K., S.D.M. Jones and M.A. Price, 1998. Methods of predicting lamb carcass composition: a review. Small Ruminant Research, Volume 29, 241-254.

Stouffer, J.R., 1988. Ultrasonics for live lamb and carcass evaluation. Methods to Improve the Lean Yield of Lamb, NC-III Technical Committee for Increased Efficiency of Sheep Production Symposium, Denver, CO, USA, 26-30.

Swatland, H.J., 1995. On-Line Evaluation of Meat, Technomic, Lancaster, PA, USA, 347 p.

Taylor, C.S., 1965. A relation between mature body weight and time taken to mature in mammals. Animal Production, Volume 7, 203-220.

Timon, V.M. and M. Bichard, 1965. Quantitative estimates of lamb carcass composition: III. Carcass measurements and a comparison of the predictive efficiency of sample joint composition, carcass specific gravity determinations and carcass measurements. Animal Production, Volume 7, 189-201.

Vangen, O., 1989. X-Ray CT for body composition. In: Kallweit, E., M. Henning and E. Groeneveld (Eds.), Application of NMR Techniques on the Body Composition of Live Animals, Elsevier, New York, USA, 91-99.

Wiese, S. and M. Wood, 2006. The European approach to carcase classification. www.texel.org.au/te-rakau/Texel_Trial_shows_retail_value.htm.

Wang, J., S.B. Heymsfield, M. Aulet, J.C. Thorton and R.N. Pierson, 1989. Body fat from body density: underwater weighing vs. dual-photon aborptiometry. American Journal of Physiology, Volume 256, 829-834.
Wells, P.N.T., 1984. Introduction to imaging technology. In: Lister, D. (Ed.), In Vivo Measurement of Body Composition in Meat Animals, Elsevier, New York, USA, 25-35.
Wood, J.D., P.B. Newman, C.A. Miles and A.V. Fisher, 1991. Video image analysis: comparisons with other novel techniques for carcass assessment. Proceedings Symposium Electronic Evaluation of Meat in Support of Value-Based Marketing, March 27-28, Purdue University, IN, USA, 145-169.

New methods for grading beef and sheep carcasses

Paul Allen
Ashtown Food Research Centre, Teagasc, Ashtown, Dublin 15, Ireland

Abstract

In the EU both beef and sheep carcasses are graded by visual assessment of their conformation and external fat cover ((S)EUROP scheme). Automatic grading machines are preferable to visual assessment because they are more consistent and producers can have more confidence in the fairness of grading results. Video Image Analysis (VIA) technology has been developed to classify carcasses according to the (S)EUROP scheme automatically. In 1999/2000 three VIA systems were installed in a beef plant in Ireland and their performance in predicting reference scores for conformation and fat class was assessed. There were relatively small differences in accuracy between the systems but all three performed better at predicting conformation scores than at predicting fat scores. In 2003 the EC regulation for beef carcass grading was changed to allow automated systems and Ireland was the first country to achieve authorisation. VIA systems have also been applied to sheep carcass grading but so far there has been little adoption by the industry. VIA systems can also accurately predict saleable yield and this can be further improved by the use of hand held cameras to take images at the quartering point. This can also yield information related to quality such as marbling and lean and fat colour. The ultimate goal should be to grade on palatability, the consumers' perception of quality, as is done in the USA and Australia.

Keywords: beef, sheep, carcass grading, image analysis

Introduction

The purpose of carcass grading is to sort carcasses into classes according to criteria related to their quality. By attaching a pricing schedule to the various classes quality-based price signals are sent to the producer providing the incentive to produce the type of carcass required by the market. For this to work effectively producers must have confidence that the grading scheme is fair, objective and consistent, otherwise they will not 'buy-in' to it, resulting in disharmony and a preference for flat-rate pricing. Grading schemes based on visual assessment, as in the (S)EUROP scheme, do not generally fulfil the criteria of being fair, objective and consistent. Automated systems that use video image analysis (VIA) technology have been developed for (S)EUROP grading. These have been shown to be at least as accurate and consistent as human graders and have been applied in several countries. Other countries such as the USA and Australia grade their carcasses after quartering when the eye muscle and overlying fat are visible and this allows for grading based on eating quality in addition to grading on yield. Hand-held VIA systems have also been applied recently in the USA to assist the human graders. Beef carcass grading schemes in the EU and USA and the application of VIA technology in these countries have been reviewed by Allen (2003). VIA systems for grading sheep carcasses have also being installed in some countries, though the take up is much slower than in beef. This paper reviews these developments.

The EU beef carcass classification scheme

European Union regulations state that beef carcasses must be classified according to their conformation and fat cover by trained classifiers (EC 1208/1981), the so-called (S)EUROP system. For conformation the classes E U R O P are used with E denoting carcasses with the best conformation. There is an option to use an extra S class for carcasses with extremely good muscle development such as double-muscled individuals. Fat cover is assessed on a five-point scale using the numbers 1-5, with 1 being the leanest. Many countries subdivide each of the categories for conformation and fat into 3 subclasses to give a 15 x 15 grid. In other countries only the most common classes are sub-divided. The classes each have descriptions and photographic standards. Classifiers, who are employed by the state, by an independent grading organisation or by the processor, are highly trained and must be regularly monitored by the responsible national organisation and retrained if necessary. Standards throughout the EU are maintained by an expert panel who visit each country on a regular basis to check that the grading is in line with the EU standards. The classification scheme is used by the EU for price reporting and market intervention purposes and by the industry for quality-based payments to producers and for carcass trading.

The main problems with visual grading are that it is viewed by many as being subjective, inconsistent and subject to influence. Despite the rigorous monitoring and control systems in place it is difficult to prove its objectivity. This can lead to a lack of confidence in the grading scheme and resistance by producers to pricing schedules with realistic price differentials for different quality grades. Flat-rate pricing can then become common removing any incentives to produce carcasses of the quality required by the market. It then becomes difficult to maintain market share in the face of increasing competition and the whole industry loses. This was the case in Ireland in the late 1990's so a collective decision was made by the industry to adopt mechanical grading as soon as possible. A comparative trial of three mechanical systems that use video image analysis (VIA) technology was carried out. This eventually led to one system being installed in all the main export plants once the EU regulations were changed to allow their use.

Development of VIA systems for beef carcass grading

The need to comply with the EC regulation by assessing conformation and external fat cover of hot carcasses and assigning classes to these based on the (S)EUROP scheme limits the possible technologies that can be applied to mechanical beef carcass grading in the EU countries. VIA technology which involves taking images of a carcass then using software to extract data such as linear measurements, volumes, angles, curvatures and colours then using these to predict the conformation and fat class is an obvious candidate. The development of this technology for beef carcass grading began in the early 1980's, and by the end of the 1990's 5 systems were available commercially. These systems were:

BCC2 was developed by SFK technology in Denmark. This uses a single colour camera, a holding frame to keep the half carcass steady and a lighting system including striped light. Three images of the outer side of the half carcass are taken while it is stationary, one with the lights on, another with them off, and a third with the striped light. The first two are subtracted from each other to take account of any variation in ambient light. The third is used to gain 3D information about the carcass from the degree of curvature of the stripes. Neural network analysis is used to predict conformation and fat classes from all the available data.

VBS2000 was developed by E + V in Germany. This also uses a single colour camera, a holding frame to keep the half carcass steady and a lighting system with striped light. Only two images are taken, as it is not considered necessary to adjust for ambient light.

MAC was developed by Normaclass in France. This uses six monochrome cameras set at different heights and viewing angles and a rotating dual holding frame. The first half carcass rests against the frame rib side out and is imaged by two of the cameras (one for the hind the other for the fore). These images are used to determine the outline of the carcass and to assess the fat on the inside.

The table then rotates 180 degrees, the first side is released and the second half carcass then comes to rest against the other frame. This side is moved into three different positions and all six cameras take images at each orientation. 3D information is gained from these different viewing angles. The MAC includes an automatic washing system.

VIAScan was developed by Meat and Livestock Australia in Australia. This system does not have a holding frame. It takes pictures with a colour camera while the half carcass is moving so it can operate at much higher speeds. The camera, lighting system and computer are all contained in a stainless steel box. The system is the most compact and is placed only about half as far from the line as the others. The hot carcass system is one of 3 systems developed by VIAScan, the other two are for grading cold carcasses after quartering and for grading cuts. Information form all three systems can be linked to improve predictions.

CVS was developed by Lacombe University in Canada (Tong *et al.*, 1997). It is very similar to VIAScan. It also operates at high line speeds as the line is not stopped. A cold cut face camera is also available for grading at the quartering point.

Performance of VIA systems

Adoption of VIA grading systems by the beef industry will depend among other things upon their effectiveness in accurately discriminating carcasses according to criteria that are related to their commercial value. Accuracy and repeatability will be important criteria in making the decision to install systems and this will be the main focus of this paper. Cost, practicality and reliability will also be important considerations. Regulatory issues will also be a factor in the timing of the adoption of the technology by the industry and this will also be discussed, particularly in relation to the EU.

Prediction of (S)EUROP conformation and fatness

The (S)EUROP system has been described above. Before commencing a review of the performance of VIA systems at predicting (S)EUROP scores for conformation and fatness it is important to note that the VIA systems have to be trained and calibrated using 'reference' scores determined by one or more human classifiers. Thus, any inaccuracy and inconsistency in the 'reference' scores is included in the error of the VIA systems. Their performance is then subsequently judged against a similar 'reference' so that the inherent inaccuracy of the 'reference' is likely to be compounded. The developers of the systems will obviously take measures to maximise the objectivity of the carcass scores used to train and calibrate their systems, such as using a panel of experienced classifiers and using only those carcasses where there is total agreement about the classification scores. It is also essential, though, to pay attention to the quality of the reference scores in trials to determine the performance of the systems.

The systems have been demonstrated to be accurate and repeatable in several trials. Madsen *et al.* (1996) reported a large trial where the BCC-2 was compared with a classifier and an inspector. The results showed that the BCC-2 was more accurate than the plant classifier for both conformation (SEP = 0.57 for BCC-2 v 0.75 for classifier) and for fat class (SEP = 0.97 for BCC-2 v 1.15 for classifier). Moreover, the BCC-2 was more repeatable that the inspector when carcasses were reclassified within 1 hour (RMSE = 0.12 v 0.51 for conformation and RMSE = 0.17 v 0.80 for fat, for BCC-2 and inspector respectively).

Two trials of the VBS2000 system were reported by Sonnichsen *et al.* (1998). In the first trial a single classifier was compared with the VIA system on 301 young bulls of three breeds. The performance differed little between breeds and was better for conformation (R^2 = 0.90, SEP = 0.93) than for fat class ($R^2 = 0.75$, SEP = 1.20). In the second trial two experienced classifiers were used as the reference and the results were improved ($R^2 = 0.91$, SEP = 0.81 for conformation and $R^2 = 0.80$, SEP = 0.91 for fat).

The MAC system was tested by the national research organisation INRA (unpublished data). For conformation 100% of young bulls and 99% of cows were classified within 2 subclasses of the reference (15-point scale). Corresponding figures for fat class were 98% and 89%.

A common feature of all the trials of VIA systems is that conformation class is predicted with greater accuracy and reputability than fat class.

A comparative trial of three VIA systems

In 1999, the first comparative trial anywhere of VIA systems was undertaken in Ireland (Allen and Finnerty, 2000). Three systems, BCC-2, VBS2000 and VIAScan were installed side by side in the Dawn Meats factory in Midleton, County Cork. The first two systems had both been developed for predicting (S)EUROP grades, but the VIAScan system had little previous experience of (S)EUROP grading. A panel of three experienced classifiers scored the carcasses using the 15 x 15 grid and agreed a consensus score when they gave different scores. This consensus score was used as the reference. It was believed that this would give a more accurate reference than using the scores of individual classifiers. A total data set of over 7,000 carcasses was divided into a calibration set (n = 4,278) and a validation set (n = 2,969). As none of the systems had previously been trained on Irish carcasses the calibration set was used by the operators of the VIA systems to derive suitable algorithms for conformation class and fat class predictions. These were then tested with the validation set. For conformation the percentage classified to within one subclass (1/3 of a class) of the reference was 96.5, 92.8 and 91.0% for VBS200, BCC-2 and VIAScan respectively. The corresponding errors (RSD) were 0.75, 0.70 and 0.80. In common with previous tests of the systems individually, the performance for fat class predictions was poorer than for conformation, with the percentage predicted to within one subclass of the reference being 74.6, 80.4 and 72.0% for VBS200, BCC-2 and VIAScan respectively, and errors of 1.38, 1.14 and 1.38 respectively.

A second validation trial was undertaken at the same factory in Ireland in 2000. All the data from the first trial (calibration and validation sets) were made available to the companies to optimise their algorithms prior to this second trial. These were then tested on over 2,000 carcasses. The performance of all three systems was again very acceptable for conformation with the percentage classified to within one subclass of the reference being 95.4, 97.0 and 94.2 for VBS200, BCC-2 and VIAScan respectively. For fat class the percentage classified to within one subclass of the reference was again lower than for conformation and was unchanged for two systems with VIAScan showing an improvement to 76.1%.

A detailed report of the two trials (Allen and Finnerty, 2000) and a shorter summary (Allen and Finnerty, 2001) have been published. The main conclusions from the two trials were as follows:

- There were relatively small differences between the three systems in their ability to predict the reference scores for conformation and fat class.
- Overall there was a small improvement in performance in the second trial, suggesting the additional data was used to improve the prediction algorithms.
- The performance of all three systems was markedly better for conformation than for fat class.
- All systems showed some biases which would need to be reduced before an authorisation trial.

Authorisation trial

In 2003, after the EC regulation had been amended to allow automated grading, an authorisation trial was carried out in Ireland on the same three systems at the same factory. Following the new EC regulation a panel of five classifiers, three from other EU countries, was used and a representative sample of 1,290 carcasses were classified independently by each classifier and by the three VIA systems. The median result of the panel was taken as the reference for each carcass and the predictions of the systems were compared with this. Scores were allocated as shown in Table 1 and all three systems passed the 600 point threshold for authorisation. They also passed the bias and slope limits shown in Table 1.

The lower penalties for errors in fat class predictions in the scoring system reflect the recognition that fat cover is more difficult to assess than conformation. The results from the comparative trails in Ireland and all other published research on VIA systems indicate that both classifiers and automated systems are less accurate at assessing fat cover than conformation. This may be because on fatter carcasses the depth of fat becomes important in addition to the total area of the carcass covered by fat and the depth is difficult to visualise. It should be noted that if classifiers have more difficulty in assessing fat cover than so will the automated systems as these are calibrated against classifiers. Although the scoring system imposes lower penalties for fat cover than for conformation the same threshold of 600 points applies to both.

Installations in Ireland

Following authorisation, the Irish beef industry selected the VBS2000 system and 24 systems were installed at all the main export plants in 2004. At the same time the Department of Agriculture and Food withdrew their classification service. Smaller plants have to train their own operatives to continue with the visual system. After a few initial teething problems all systems are working well. Early claims that the systems underscored good conformation carcasses were shown to be unfounded. There are now much fewer complaints about grades than before their introduction and the issue has ceased to be a major source of discontent. Some companies have indicated their interest in having saleable yield predictions but as yet there is no move towards a national scheme based on yield. In the absence of this each company will probably have its own calibrations based on their main cutting specifications.

Table 1. Scoring system for authorisation of automated grading equipment.

	Conformation	Fat cover
No error	10	10
Error of 1 subclass	6	6
Error of 2 subclasses	-9	0
Error of 3 subclasses	-27	-13
Error of more than 3 subclasses	-48	-30
Total score	>600	>600
Bias	± 0.3	± 0.60
Slope of the regression line	1 ± 0.15	1 ± 0.30

Prediction of saleable yield

Conformation and fatness as assessed in the (S)EUROP scheme are both related to the commercial value of a carcass, but this relationship is mostly, though not totally, due to their effect on the saleable yield. When the EU scheme was devised there were no satisfactory instrumental methods of measuring yield on line so the best tool available was a visual assessment of conformation and fat cover. No attempt was made to convert the different classes into yield percentages, as there was no standard definition of yield. If VIA systems are able to predict saleable yield and other quality related traits more accurately than the (S)EUROP grading then they would have an additional value to the industry.

Borgaard *et al.* (1996) showed that the BCC-2 was more accurate than a classifier in predicting the percentage saleable yield (SEP = 1.34 v 1.63), the percentage hindquarter (SEP = 1.01 vs. 1.26) and the ribeye area (SEP = 5.8 vs. 6.7). VIAScan was shown to be more accurate at predicting saleable yield than the existing grading system that used weight and a fat depth for three out of our types of carcasses (Ferguson *et al.*, 1995). Standard errors for the VIAScan were between 0.98 and 1.52%. Sonnichsen *et al.* (1998) reported a slightly higher SEP of 1.8% for predicting the saleable yield of 301 young bulls of three breeds. However, it is not meaningful to compare the results of different trials due to differences in the variability of the samples and in the specification of saleable yield.

The first Irish trial is the only comparative trial to have been conducted (Allen and Finnerty, 2000). A sample of nearly 393 steer half carcasses were boned out and trimmed to a commercial specification. Roughly two thirds of these were used to calibrate the three systems and the rest (n = 133) were used for validation. There were only small differences between the three systems in their ability to predict saleable yield, with RSD's between 1.12 and 1.20%. Surprisingly though, the classification scores plus carcass weight were equally accurate (RSD = 1.21). This may have been due to the fact that a panel of three classifiers was used and the consensus scores were probably more reliable than those of individual classifiers as used in other trials. The fact that the specification did not involve heavy trimming of fat may also have been a factor. In fact the percentage saleable yield actually increased as fat class increased, the opposite to what would be expected. Primal yield was therefore calculated by excluding the trim and the flank. In this case the primal yield followed the expected trend, decreasing with increasing fatness. However, the VIA systems were less accurate than the classification scores and weight at predicting primal yield (RSD = 1.44 vs. 1.50 – 1.56).

Future possibilities for beef carcass grading

In the USA and in Australia, grading is done after chilling and quartering. This allows the eye muscle and overlying fat to be seen. In the USDA scheme carcasses are graded based on estimated yield, using factors such as weight, eye muscle area and fat depth, and on quality using factors such as ossification, marbling fat, lean colour and fat colour. Handheld VIA systems have been developed to do at least part of this automatically. Cannell *et al.* (1999) showed that the dual component VIAScan was superior to the online grader at all three trimming specifications tested, but was less accurate than an expert grader working without time constraints. However, when the VIAScan ribeye area was combined with the expert grader's estimates of the other yield grade factors the percentage of the variation in yield at all trim levels was greater than for the expert grader alone. This illustrates the potential of VIA systems in augmenting the grader and allowing more time for the other factors to be assessed by the grader. Interestingly, the percentage variation explained increased as the trim level increased from commodity to closely trimmed for all models, but there was a further increase as the trim increased to very closely trimmed only for the VIAScan system alone. At this level of trim the advantage of augmenting the grader compared to using the VIAScan alone was marginal (75 vs. 71%). Similar results were found using the dual component CVS system (Cannell *et al.*, 2002). The

CVS system was far superior at predicting closely trimmed wholesale yield than the online grader (% variation explained = 64 vs. 39) but almost as accurate as the expert grader working offline when used alone or to augment the expert grader. Shackelford *et al.* (2003) conducted trials in two commercial beef processing facilities of the Meat Animal Research Centre (MARC) system. Prediction equations that included image analysis variables and hot carcass weight accounted for 90% of the variation in calculated yield grade compared to 73% for the online graders. MARC data plus carcass weight also accounted for 88, 90, 88 and 76% of the variation in eye muscle area, preliminary yield grade, adjusted preliminary yield grade and marbling score respectively. The authors concluded that the system could be used to determine yield grade on-line but is not accurate enough to be used alone for predicting marbling score.

Adoption of handheld VIA cameras within the EU to take images at the quartering point to augment the on-line VIA data could potentially improve the accuracy of yield predictions. It would also give information related to the potential eating quality of carcasses, such as marbling fat and fat and lean colour that could be used by the industry to apply quality based grading schemes but also could feed in to the breeding schemes to improve yield and quality in the future.

It is important to remember that quality to the consumer really means eating quality, mainly tenderness, juiciness and flavour. The current EU grading system has little to do with this. With increasing competition internationally an industry that doesn't seek to satisfy the consumer is unlikely to prosper. A change towards grading and quality payments based on consumers' expectations which will feed in to breeding schemes and production methods is the way forward.

Grading based on eating quality depends on being able to measure or predict eating quality. As yet there are no very reliable methods to measure eating quality on-line or shortly after slaughter. Belt grill cooking and shear force determination of *longissimus* steaks cut at 1-2 days post mortem has been shown to give repeatable and reliable predictions of sensory panel ratings of tenderness (Shackelford *et al.*, 1999) and Vote *et al.* (2003) concluded that on-line measurements by a CVS VIA system equipped with a BeefCam were useful for predicting the tenderness of beef *longissimus* muscle steaks. Wheeler *et al.* (2003) showed that beef colour, either measured by the BeefCam VIA system or by colorimetry, were sufficiently accurate at predicting tenderness. More recently, a device based on NIR spectroscopy has been demonstrated to be useful for sorting carcasses into acceptable and unacceptable categories (Shackelford *et al.*, 2005). In Australia the approach taken by Meat and Livestock Australia is to predict eating quality from a number of live animal and post slaughter factors using a model, the MSA palatability model, developed over 10 years and based on over 70,000 samples tasted by large numbers of consumers. This model predicts palatability scores for all the main cuts in the carcass cooked by a number of different methods. This approach recognises the reality of the variability in eating quality between different muscles within a carcass and the interaction between muscle and cooking method. At the Ashtown Food research Centre we are currently testing this model on Irish beef and Irish consumers.

Sheep carcass grading

In the EU sheep carcasses are graded visually according to their conformation and fatness in a similar fashion to beef carcasses. Reflecting the considerable variation in slaughter weights between different regions and production systems there are different scales for carcasses above and below 15kg. The criticisms that are levelled at the beef carcass visual grading can also be applied to the sheep carcass scheme. The VIA technology has also been applied to sheep carcass grading. VIA has been shown to be better than visual grading for predicting meat yield (Stanford *et al.*, 1998, Brady *et al.*, 2003 and Hopkins *et al.*, 2004). However, as yet there has been little adoption by the industry, possibly because of the high cost of the equipment and the speed of the lines.

Other technologies for carcass grading

Other technologies that have either been applied to carcass grading or have potential for this application include whole body electrical conductivity (TOBEC), bio-electrical impedance analysis (BIA), ultrasound and magnetic inductance technology. Berg *et al.* (1997) showed that TOBEC could predict the fat and lean content of lamb carcasses with sufficient accuracy for grading applications but the technology is not applicable to beef carcass grading due to size limitations. BIA has also been shown to be a useful predictor of composition in both beef (Bohuslavek *et al.*, 2004) and lamb (Berg and Marchello, 1994) carcasses and could be used to augment the VIA data as BIA measures internal rather than external fat, but there are practical difficulties related to the placement of electrodes on the carcass. Ultrasound has been successfully applied to the measurement of fat and muscle depths in pig carcasses but these measurements are less repeatable and have poorer correlation with overall composition of beef carcasses. Reasons for this include the uneven nature of the subcutaneous fat cover on beef and sheep carcasses, which is exacerbated by hide removal and the fact that the subcutaneous fat depot is a smaller proportion of the total carcass fat than in pork carcasses (Kempster *et al.*, 1982). Finally, X-Ray CT and MRI have been shown to be highly accurate at estimating carcass composition (Dobrowolski *et al.*, 2004; Collewet *et al.*, 2005; Romvari *et al.*, 2006) but have little prospect of being used on-line due to cost and technical issues.

References

Allen, P., 2003. Beef Carcass Grading in (S)EUROPe and U.S.A. – The Prospects for Using VIA Systems. Brazilian Journal of Food Technology, Special Issue, 96-101.

Allen, P. and N. Finnerty, 2000. Objective beef carcass classification – A report of a trial of three VIA classification systems. Teagasc, Dublin, Ireland.

Allen, P. and N. Finnerty, 2001. Mechanical grading of beef carcasses. Teagasc End of Project Report, No, 45, Teagasc, Dublin, Ireland.

Berg, E.P. and M.J. Marchello, 1994. Bioelectrical impedance analysis for the prediction of fat-free mass in lambs and lamb carcasses. Journal of Animal Science, Volume 72, 322-329.

Berg, E.P., Neary, M.K., Forrest, J.C., Thomas, D.L. and Kaufmann, R.G., 1997. Evaluation of electronic technology to assess lamb carcass composition. Journal of Animal Science, Volume 75, 2433-2444.

Bohuslavek, Z., W. Branscheid and C. Augustini, 2004. Prediction of beef grades by means of the bioelectrical impedance analysis (BIA). Fleischwirtschaft, Volume 84, 108-111.

Borggaard, C., N.T. Madsen and H.H. Thodberg, 1996. In-line image analysis in the slaughter industry, illustrated by beef carcass classification. Meat Science, Volume 43, S151-S163.

Brady, A.S., K.E. Belk, S.B. LeValley, N.L. Dalsted, J.A Scanga, J.D. Tatum and G.C. Smith, 2003. An evaluation of the lamb vision system as a predictor of lamb carcass red meat yield percentage. Journal of Animal Science, Volume 81, 1488-1498.

Cannell, R.C., K.E Belk, T.D. Tatum, J.W. Wise, P.L. Chapman, J.A. Scanga and G.C. Smith, 2002. Online evaluation of a commercial video image analysis system (Computer Vision System) to predict beef carcass red meat yield and for augmenting the assignment of USDA yield grades. Journal of Animal Science, Volume 80, 1195-1201.

Cannell, R.C., J.D. Tatum, K.E. Belk, J.W. Wise, R.P. Clayton and G.C. Smith, 1999. Dual-component video image analysis system (VIAScan^{TM}) as a predictor of beef red meat yield percentage and for augmenting application of USDA yield grades. Journal of Animal Science, Volume 77, 2942-2950.

Collewet, G., P. Bogner, P. Allen, H. Busk, A. Dobrowolski, E. Olsen and A. Davenel, 2005. Determination of the lean meat percentage of pig carcasses using magnetic resonance imaging. Meat Science, Volume 70, 563-572.

Dobrowolski, A., R. Romvari, P. Allen, W. Brandscheid and P. Horn, 2004. X-Ray computed tomography as possible reference for the pig carcass evaluation. (In German with summary in English). Fleischwirtschaft, Volume 84, 109-112.

Ferguson, D.M., J.M. Thompson, D. Barrett-Lennard and B. Sorensen, 1995. Prediction of beef carcass yield using whole carcass VIAScan. Proceedings 41st ICoMST, San Antonio, USA, Paper B16, 183-184.

Hopkins, D.L., E. Safari, J.M. Thompson and C.R. Smith, 2004. Video image analysis in the Australian meat industry - precision and accuracy of predicting lean meat yield in lamb carcasses. Meat Science, 2004, Volume 67, 269-274.

Kempster, A.J., A. Cuthbertson and G. Harrington, 1982. Carcase Evaluation in Livestock Breeding, Production and Marketing. Granada Publishing Ltd., London, UK.

Madsen, N.T., H.H. Thodberg, T. Flig and E. Ovesen, 1996. BCC-2 for objective beef carcass classification and prediction of carcass composition. Proceedings 42nd Annual Reciprocal Meat Conference, Lillehammer, Norway, Paper G-6, 244-245.

Romvari, R., A. Dobrowolski, I. Repa, P. Allen, E. Olsen, A. Szabo and P. Horn, 2006. Development of a computed tomographic calibration method for the determination of lean meat content in pig carcasses. Acta Veterinaria Hungarica, Volume 54, 1-10.

Shackelford, S.D., T.L. Wheeler and M. Koohmaraie, 1999. Evaluation of slice shear force as an objective method of assessing beef longissimus tenderness. Journal of Animal Science, Volume 77, 2693-2699.

Shackelford, S.D., T.L. Wheeler and M. Koohmaraie, 2003. On-line prediction of yield grade, longissimus muscle area, preliminary yield grade, adjusted preliminary yield grade and marbling score using the MARC beef carcass image analysis system. Journal of Animal Science, Volume 81, 150-155.

Shackelford, S.D., T.L. Wheeler and M. Koohmaraie, 2005. On-line classification of US Select beef carcasses for longissimus tenderness using visible and near-infrared reflectance spectroscopy. Meat Science, Volume 69, 409-415.

Sonnichsen, M., C. Augustini, A. Dobrowolski and W. Brandscheid, 1998. Objective classification of beef carcasses and prediction of carcass composition by video image analysis, Proceedings 44th ICoMST, Barcelona, Spain, Paper C59, 938-939.

Stanford, K., R.J. Richmond, S.D.M. Jones, W.M. Robertson and A.J. Gordon, 1998. Video image analysis for on-line classification of lamb carcasses. Animal Science, Volume 67, 311-316.

Tong, A.K.W., R.J. Richmond, S.D.M. Jones, D.J. Robinson, B.P. Chabot, S.M. Zawadski, W.M. Robertson, X. Li and T. Liu, 1997. Development of the Lacombe computer vision system (Lacombe CVS) for beef carcass grading. Agriculture and Agri-food Canada, Lacombe Research Centre, AB, Canada.

Vote, D.J., K.E. Belk, J.D. Tatum, J.A. Scanga and G.C. Smith, 2003. Online prediction of beef tenderness using a computer vision system equipped with a BeefCam module. Journal of Animal Science, Volume 81, 457-465.

Wheeler, T.L., D. Vote, J.M. Leheska, S.D. Shackelford, K.E. Belk, D.M. Wulf, B.L. Gwartney and M. Koohmaraie, 2003. The efficacy of three objective systems for identifying beef cuts that can be guaranteed tender. Journal of Animal Science, Volume 80, 3315-3327.

Beef carcass grading and meat quality measurements in different countries and how ICAR is going to use such information

Laurent Journaux
UNCEIA, MNE, 149 Rue deBercy, 75595, Paris Cdex 12, France

Abstract

For beef cattle, records on carcass trait include essentially global traits like carcass weight, carcass conformation, and in USA on ultrasonic measurement. This type of record is developing very fast on commercial carcass. In the guidelines of the international organisations like ICAR and BIF, you find in additional recommendations to record Warner Bratzler shear force. But there is no recommendation concerning other traits related to meat quality like pH, fibres. The major explanation is that these measures are difficult to obtain and generally expensive. So it's difficult to find their cost return in a breeding scheme, even if the industry is more and more interested by a selection on this family of traits. The new developments in genomics will offer a new way to integrate this type of information in beef breeding schemes by using DNA tests.

Keywords: carcass grading, meat quality, ICAR, BIF

Introduction

The International Committee for Animal Recording (ICAR) is the world-wide organization for the standardization of animal recording and productivity evaluation. Its aim is to promote improvement of farm animal recording and evaluation through the formulation of definitions and standards for the measurement of traits of economic importance. Together with the definitions and standards ICAR establishes specific guidelines for the purpose of identifying animals, the registration of their parentage, recording their performance and their evaluation, and publish the findings (www.icar.org). To achieve these goals, ICAR is structured with several permanent sub-committee and non permanent working group. These working groups have to perform the actions defined in a term of reference approved by the ICAR general Assembly.

Since 1986, a beef working group is continually in the process of maintaining and updating universal guidelines for use in recording beef performance data world wide (Rehben, 1990). The objective is also to promote the application of these guidelines in ICAR member and non-member countries. The primary objective of the guidelines is to facilitate easier exchange of data through international harmonisation of recording procedures (and trait descriptions) and making global genetic evaluation easier and more meaningful. The understanding of and the reliability of recorded beef data will so be greatly improved by describing the recording methodology clearly and having this readily available internationally. A major objective of ICAR is to facilitate and lead the development (and sustained practice) of a joint international genetic evaluation system for beef traits. The Working Group also involves itself in a regular survey of recording practices and statistics regarding beef cattle globally. A major new objective is to include countries classified as developing economies in the activities of the Working Group and to assist in the establishment and maintenance of recording systems for beef cattle in these countries. Due to the developments in global genetic evaluations for beef cattle the Working Group has also set a goal for itself to encourage more presentations at the ICAR and INTERBULL (sub committee of ICAR in charge of international genetic evaluations) technical sessions related to this subject. So, the Beef Working Group of ICAR is very interesting to have state of art of the new measurements available to describe meat carcass quality.

ICAR recommendation for beef performance recording

Basements of ICAR recommendation

The current ICAR guideline aims to provide guidelines for the relevant matters which must be undertaken in the routine execution of beef recording schemes (ICAR, 2005).

Beef production is predominantly based on specialised beef breeds that use natural mating, the rearing of calves by their mothers and the finishing of the young animals in specialized finishing units. On the other hand, dual purpose and dairy breeds that mainly use artificial insemination and separate the young calf from the mother immediately after birth, also contribute significantly to beef production in many countries. Therefore, the guideline aims to provide for the recording of all cattle used for meat production. The ICAR surveys (Journaux *et al.*, 2002; Journaux, 2004; Simianer *et al.*, 2001) clearly indicated two main beef recording traditions. The European type approaches on the one hand and North American type approaches, as represented by the Beef Improvement Federation (BIF, www.beefimprovement.org), on the other hand. The differences between them can in the main be traced back to substantial differences in consumers demand impacting the pricing system and consequently the selection objectives and also the significant differences in the production environment and in particular herd sizes.

The ICAR guideline aims to combine recording standards of all regions in as much as this is possible. However, overall uniformity can not be fully accomplished. The guideline recommends basic procedures. However, there will be situations where national organizations will develop more refined procedures that are more suitable for their members. The ICAR recommendations describe several traits linked to the carcass and meat quality that can be recorded with a standardized method.

The first group concerns the *description of the carcass.*

Carcass weight

Carcass weights are unaffected by variation in shrinking and therefore apart from the scaling effect - show less variation than live weights. Compared to live weights they relate more to meat yield and to the consumers endpoint. Calculation of net gain is based on slaughter weight. Typically, carcass weights are collected by commercial abattoirs; additionally experimental abattoirs come into consideration. Carcass weights should be collected consistently to ensure an informative data analysis.

Usually, carcass weight is defined by appropriate national legislation which clearly specifies which parts of the carcass are to be removed prior to taking the weight. In the case of no legal definition, carcass weight should be defined as the hot weight of both half carcasses after removal of skin, bled and eviscerated and after removal of external genitalia, the limbs at the carpus and tarsus, head, tail, kidneys and kidney fats and the udder. Preferably the unit of measurement should be metric to the nearest of 500 grams.

Carcass grade

Carcass grades significantly affect the market value of the carcass. Therefore they form a trait with big economic impact and should be used for the analysis of progeny productivity. Grading mostly is done according to national standards that frequently are based on appropriate legislation.

However, according to different market demands, national grading schemes frequently target different objectives and therefore are composed of different traits. On a global level there are two predominant types of grading schemes:

- USDA grading scheme includes the following components: Class (steer, bullock, bull, heifer, cow), Maturity (eat colour and texture of lean meat), Quality grade in 8 levels (Prime, Choice, Select, Standard, Commercial, Utility, Cutter, Canner) including evaluation of marbling and firmness; Yield grade (in combining external fat, kidney, pelvic and heart fat, ribeye area, carcass weight). It is recommended that a highly trained and certified person be used to assess quality grade factors when collecting carcass data.
- EU grading scheme includes the following components: Class (calf, young bull (=bullock), bull, steer, heifer, cow), Conformation grade in 6 levels (S-E-U-R-O-P for the best to the worst situation), Fat grade in 5 levels (1-2-3-4-5 from the learner to the fatter), and for veal meat the colour (1 to 5, from white to red) (http://www.office-elevage.fr/publications/classement/classement.htm).

As a consequence meat reports are almost incomparable across big market regions like e.g. North America, Europe and other continents. Therefore the grading system should be clearly indicated on reports provided for use outside the country where the grading scheme is applied. In order to provide useful information that might be used outside the market region, it is recommended additionally to record each of natural components forming the grade.

Dressing percentage

Dressing percentage describes the percent ratio between carcass weight and the live weight taken immediately before slaughter. Although dressing percentage mainly is used for the estimation of carcass weights of live animals, it provides additional information on the animal's type even if carcass weight is measured directly. A scale that measures in increments of 1 kg or 2 lb, or less, should be used for taking the live weight immediately before slaughter.

As live weight is largely influenced by shrinking, dressing percentage should account for this effect, by standardisation of the live weight to 12-hours shrinking time. The correction factors should apply in the special production environment of the animals. Dressing percentage should be described as percentage with 1 decimal place.

Meat yield

Meat yield means the percentage of lean meat in the beef carcass as obtained by dissection. However, - with regard to high costs arising from carcass dissection - meat yield frequently is estimated on the base of surrogate traits, which can be easily measured in the course of the slaughter process. In some areas meat yield refers to the whole lean meat contained in the carcass, whereas other regions account for specified retail cuts forming the most evident part of the carcass value. Meat yield should be described as percentage with 1 decimal place.

Some areas apply yield grades rather than meat yield itself; e.g. the USDA yield grade is a numerical score from 1 to 5 expressed as a whole number. It represents the yield of the boneless, closely trimmed retail cuts from the round, loin, rib and chuck. These cuts represent about 75% of the carcass weight and about 90% of the carcass value.

Another group of traits is more related directly to the *meat quality*. First of all we need to define the meat quality. In broader terms, quality refers to palatability, appearance, nutritional value and food safety. In practice, quality refers to the overall appearance and palatability of the edible portion of the carcass.

Quality can be determined by evaluation of animal maturity, tenderness, subcutaneous fat, intramuscular fat (marbling), meat colour, fat colour, firmness of meat (lean) and texture of meat. Factors such as juiciness, flavour, aroma and undesirable flavours (off-flavours), are also quality traits, but can only be assessed through sensory taste panels and are therefore rarely recorded and evaluated.

Meat quality can be assessed on the basis of a subjective score (including e.g. a marbling score), through taste panels, or by using technical devices to measure the meat colour, tenderness, intramuscular fat, physiological parameters like the pH at different points of time, etc.

Meat quality can probably be defined as comprising four aspects of importance: Visual quality (factors evaluated in classifying carcasses and/or factors that affect consumers decisions when purchasing meat - e.g. subcutaneous fat cover, bone content and meat and fat colour); Eating quality (tenderness, juiciness, odour and flavour intensity of the cooked product); Nutritional quality (proportions of proteins, vitamins and minerals relative to energy density); Safety (negligible risk from food-borne illness or poisoning and absence of drug, chemical, antibiotic or hormone residues). ICAR recommendations focus on visual quality and eating quality (palatability).

Maturity

Maturity can be defined as an estimation of the physiological age of the carcass, which can be determined by evaluating the size, shape, and ossification of the bones and cartilage, the number of permanent incisors and the colour and texture of the lean. Alternatively the chronological age of the animal may be used, although physiological and chronological age are not necessarily the same.

Where the chronological age of the animal is unknown, maturity score is a useful unit of measurement. Maturity is usually classified according to the percentage ossification of the cartilage of thoracic buttons. In case maturity scoring a specific scale is adapted.

Marbling

Marbling can be defined as the flecks of fat in the lean. Marbling is usually evaluated visually in the rib-eye muscle, which is exposed between the 12th and 13th ribs. Marbling contributes to meat tenderness and is also associated with the palatability traits of juiciness and flavour.

Marbling is usually assessed by classification (e.g. 9 degrees of marbling, ranging from Practically Devoid to Abundant) related to the estimated percentage of intramuscular fat. Marbling scores and intramuscular fat percentages are specific to carcass assessments performed in North America and are not necessarily applicable to other countries. As a consequence, marbling should be recorded according to BIF standards, where each degree of marbling is divided into tenths within each degree of marbling (Hohenboken, 2002).

Colour firmness and texture of lean

Colour of the rib eye muscle is used as an additional indicator of maturity or physiological age. The visual appeal of beef at the retail counter is highly dependent on desirable colour. Dark cutters are carcasses that produce lean tissue that is dark red to almost black and often result from cattle that have been stressed prior to slaughter. Dark cutters are safe to eat and their palatability is not seriously affected. However, the colour reduces consumer acceptability and lowers carcass value dramatically.

Firmness of lean refers to the relative firmness or softness of the rib-eye muscle, whereas texture of lean refers to the apparent fineness or coarseness of muscle fibres within the rib-eye muscle.

Colour, firmness, and texture of lean are widely used in North America, and are not necessarily applicable to other countries. Accordingly, those traits should be recorded according to the BIF standards (Hohenboken, 2002).

Warner Bratzler shear force

More direct measures of palatability than quality grade include Warner Bratzler shear tests for tenderness assessment, and trained sensory panel evaluation for tenderness, flavour, and juiciness. However, cost and availability will restrict usage of these alternative methods.

An initiative to standardize the protocol for Warner Bratzler shear force determinations was identified at the National Beef Tenderness Plan Conference in April, 1994. The purpose of this protocol is to facilitate consistent collection of Warner Bratzler shear force determinations across institutions for comparative evaluation. These data can be used in progeny testing and in the development of carcass breeding values to improve meat tenderness. Any institution abiding by these guidelines can be certified to collect Warner Bratzler shear force determinations for the beef industry.

These specific recommendations do of the Standardized Warner Bratzler shear force a suitable trait for procedures for sire evaluation. They include elements concerning the condition of conversion of live animals to carcasses, sample preparation (choose of the muscle, cut of this muscle and cocking conditions), Core preparation. In addition, of institutions that perform Warner Bratzler shear force measurement must be certified.

A last family of traits, concerning body composition can be recorded in live using ultrasonic measurements.

Ultrasound measurements

Real time ultrasound imaging equipment to record carcass characteristics in live animals for livestock improvement programs has been in use for more than two decades. Its usefulness in beef cattle has been well demonstrated (Wilson *et al.*, 1998).

Ultrasound scanning has been used since the late 1980s in many beef cattle breeding programs to overcome the inherent difficulty of recording carcass data from progeny tests under extensive production systems and in performance test situations where access to carcass information is not possible. A number of genetic evaluation programs have now included scan data in their routine analysis. The application of ultrasound is highly technical and requires: the use of sophisticated equipment, strict adherence to proper equipment calibration, proper animal preparation, adherence to a standard scanning protocol, adherence to a standard image interpretation protocol, suitable animal handling facilities.

It is important for genetic evaluation that animals are allowed to express their inherent genetic potential. As fat measurements are directly related to the nutritional state of the animals it is essential to record only groups of animals which are on a reasonable level of nutrition. Otherwise too many animals will be recorded with minimum fat levels and no intramuscular fat, thus generating information of little value since the true genetic potential will not have been expressed. Such data is useless for genetic evaluation where the intention is to identify genetic differences.

As ultrasound measurements are used to provide an insight into a number of carcass characteristics and to a limited extent into meat quality, the most valuable records will come from young animals undergoing selection for breeding and on which no direct carcass information can be collected. Yearling bulls and yearling heifers are the most obvious animals to scan. In many commercial production systems a progeny test through steers or bulls is also possible.

In summary scanning can provide useful information for the estimation of carcass EBVs or EPDs using records from yearling bulls, yearling heifers or groups of progeny fed for slaughter. The most common age window for young breeding stock is between 320 to 500 days. It may vary depending on production system. The development of body composition EBVs or EPDs requires that scanned animals be associated with a well-defined contemporary group. For animals scanned on the farm of birth a contemporary group is comprised of all animals of the same sex that are reared and managed together. A 60 days birth window is recommended. Where herd sizes are small and calving season extended the contemporary group may cover a longer birth season window. A typical contemporary group definition would include herd code, birth season, weaning group (date, location, and management), operator (if scanned by more than one operator) and scanning group (date, location, and management). For animals scanned at a central station test, the contemporary group should include animals from the same sex born within 60-90 days age window and the same test end. The herd of origin and other birth and weaning group information may also be included.

The practise of harvesting/slaughtering animals from groups when they reach market target weight reduces the management group size as records from animals slaughtered on different days and in particular in different abattoirs should not be directly compared. Scanning for carcass traits of all animals prior to the first selection of any animals to be slaughtered will provide a basis for direct comparison of all animals in the group.

Real time ultrasound scanning for subcutaneous fat can also be used to determine market suitability of commercial slaughter animals. However, scanning of animals that have reached target market specifications should not be compared with the use of the same technology for performance recording purposes. Special care must be taken to avoid any bias in the mean of the observations. Such a bias could have severe financial implication if animals are slaughtered and found to be outside market specifications. For the purpose of genetic evaluation a consistent bias will be part of the management group effect and will not affect the accuracy of genetic evaluation.

A number of real time ultrasound recording devices are on the market. Most of them have been developed for human health or veterinary purposes (e.g. pregnancy testing). The small transducer used for medical purpose is of limited use for scanning of carcass characteristics and so special transducers are required when scanning for carcass traits.

Efficient ultrasound scanning of large groups of animals requires well designed yards, races and chutes to hold the animals in a stress free and secure manner and release them as soon as all necessary information has been recorded. The operator should insure that the cattle handling facilities for scanning are adequate in respect of health and safety considerations before he commences scanning. A squeeze chute with fold-down side panels is best for scanning beef cattle. A shaded area is required to allow the operator a good view of the monitor, as direct sunlight will make it difficult to see the images on the screen. Therefore the chute should be located under a roof that can block direct sunlight and provide protection from rain or other inclement weather conditions. A clean and grounded power signal is required at the chute-side. It is best if the electrical circuit is a dedicated line to the chute, free from the interference of other electrical equipment such as motors, etc. Most ultrasound equipment does not operate efficiently and accurately when the ambient air temperature falls below 8 °C or 45 °F. The breeder should make provisions to keep the facility heated in these situations. The operator should provide some portable supplemental heating systems to keep the ultrasound equipment warm if required.

Animals should be cleaned and clipped particularly in winter or early spring when their hair is too long to get quality images. The requirement for clipping is even higher if scanning is used to determine intramuscular fat percentage (IMF%) as the lack of complete contact between the ultrasound transducer and the animals skin can have a direct effect on the predicted IMF%. In general the length of hair coat should be no more than 1.5 cm or 0.5 inches. Prior to scanning a liquid, commonly vegetable oil, should to be applied to the scan site to provide maximum contact between transducer and skin. The temperature of the oil applied to the skin should be above 20 °C for best results. This might require a warm water bath for the bottle containing the oil during times of lower temperatures. Wet animals can be scanned successfully as the water can easily be removed from the scan area. For the scanning of eye muscle area a curved guide or offset made from super-flap will help and will allow a curved image to be recorded without the need to apply excessive pressure to maintain good contact as this would result in distortions of the muscle or fat measurements resulting.

Real Time ultrasound imaging has so far been used for the measurement of subcutaneous fat cover as well as for eye muscle area and muscle depth and the intramuscular fat percent in the *longissimus dorsi* muscle.

Rump fat thickness or P8 scan is an indicator of fatness and can be used to improve the overall accuracy of external fat measurements. This measurement can be particularly beneficial when scanning leaner animals such as yearling bulls. For measurements the ultrasound transducer is aligned directly between the hook- and pin bones without a standoff guide to collect this image. Rump fat thickness is measured at the apex of the *biceps femoris* muscle. The site is located at the perpendicular intersection of the line from the high bone (3rd sacral vertebra) with a line from the inside of the pin bone (*tuber ischii*). Rump fat thickness should be reported to the nearest millimetre or 1/25 of an inch. Operators may report to a higher degree of accuracy if desired.

The selection of the site for rib fat and eye muscle depth or area may coincide with the traditional quartering site of beef carcasses in the country. In general the records on different sites are genetically highly correlated, however they might show different variation and are more or less easy to record as different muscles might interfere. A common site assessed in a number of countries (e.g. Australia, Canada, New Zealand, USA) is located ¾ of the distance from the medial to the dorsal end of the *longissimus dorsi* muscle at a lateral point between the 12th and 13th rib. Rib fat thickness will be reported to the nearest millimetre or 1/25 of an inch. As with rump fat thickness recordings may be reported to a higher degree of accuracy. Rib and rump fat thickness are well correlated (genetic correlation exceeding 0.70) with rib fat commonly having a lower mean. However, interactions between breed, management system and environment exist.

Carcass rib eye usually is measured between the 12th and 13th ribs of the ribbed carcass. The ultrasound rib eye measurement commonly is made from the same image used to measure 12th-13th rib fat thickness.

Eye Muscle Area/Eye Muscle Depth is measured as the cross sectional area of the *longissimus dorsi* muscle. Care should be taken not to include other muscles that occur at this site. Similarly the image should be taken between the ribs not over a rib as the latter will cause distortion. The presence of well-defined intercostal muscles under the *longissimus dorsi* is an indication that the transducer is properly aligned directly between the 12th and 13th rib for this measurement.

Intramuscular fat percent or marbling is an important meat quality characteristic in certain high priced markets, because consumers equate it with outstanding eating quality. The carcass benchmark for intramuscular fat is the chemical extraction of all fat from a meat sample taken as a slice off the *longissimus dorsi* muscle. Most analytical software for IMF% use a longitudinal image in the region of the 11th, 12th and 13th ribs approximately 2/3rd of the distance from the medial to the dorsal end of the *longissimus dorsi* muscle. In experiments it has been demonstrated that the correlation between a longitudinal sample and a cross sectional sample is very high. Research has shown that variation between images on the same side is larger than variation within an image selecting different but overlapping areas for the analysis.

The IMF% trait is the most difficult of all ultrasound traits to measure accurately. Equipment calibration, animal preparation, electrical power signal noise, existence of atmospheric radio waves, and transducer-animal contact are just some of the factors that can influence the measurement accuracy. Therefore it is strongly recommended that the IMF% result be reported as the average of at least three images and even better, the average of five images to increase the accuracy. Most machines do not provide a direct measure of IMF% and thus there is a requirement for specialised PC software. An image frame from the ultrasound scan is digitised and analysed on a computer. Such analysis software is normally designed specifically for a particular ultrasound machine (Hassen *et al.*, 2001). France preferred to develop a measure by go through the muscle of the animal more adapted to the lean French breeds as Charolais limousine or Blonde d'Aquitaine. We choose to use the VOS (velocity of sound) material developed by Miles and collaborators in UK at the beginning of the 80's (Fursey *et al.*, 1991; Renand and Fisher, 1997).

The record of carcass and meat quality traits around the world

The ICAR guidelines describe today a large number of traits to record on carcass or on meat quality, but the development of this type of record for genetic evaluation purpose is not very important (Table 1 to 3) but is in progress. We can specially notice a recent large development of record on field in abattoir. For example, in France in 2005, 185 402 carcass weights and carcass conformations were recorded to be integrating in a national genetic evaluation which will be available in 2008 (Fouilloux *et al.,* 2002; Journaux *et al.*, 2005).

Only the more common and global traits like carcass weight, conformation and fat grade are recorded on a large scale. Germany was a pioneer in the record of field carcass weight in complement to the milk recoding system to obtain breeding values for milk and beef traits used in the selection of Simmental (Flekvieh) as a dual purpose breed (Engellandt *et al.*, 1999). USA have the specificity to record a lot of information using ultrasonic measurements.

The specific parameters of meat quality (tenderness and other) are never recorded in the frame of routine beef improvement scheme. The only situation where this type of data is recorded is the experimental scheme especially in relation with the development of genomic studies. The hope is to use directly genomic information to predict meat quality without specific and expensive measurements like Warner Bratzler shear force or panel testing. This type of experimental design is existing in France (Renand *et al.*, 2006), Australia and North America (Arthur and Herd, 2006).

Table 1. Live measurement on field or on performance or progeny test.

Country	Traits[1]			
Country	EMAs	IMFs	BFs	VOS
AUS	35 640	32 120	35 868	
FRA				2 100
GBR	2 500		2 500	
NZL			*	
USA	108 602	105 481		

[1]EMAs: Eye muscle area or rib eye area by ultrasonic scanning; IMFs: intra muscular fat by ultrasonic scanning; BFs: Back fat by ultrasonic scanning; VOS: Fat percentage estimated by use of the Velocity of Sound material.
*Record of data exist but we haven't the figure.

Table 2. Carcass measurement in organized progeny test.

Country	Traits[1]										
	CW	CC	Fat	IMF	BF	EMA	M%	MS	CM	MC [2]	KO
AUS	300			300		300					
CAN	285	285	285		285	285					
DEU	2 941	2 360	2 344				129	1 955			
FRA	2 100	2 100	2 100						2 100	800	
IRL	261	261									261
SVN	374	374	374			187	187			187	

[1]CW: Carcass weight; CF: Carcass conformation (EU scale); Fat: Fat grade with EU scale; IMF: Intra muscular fat; BF: Back fat fitness; EMA: Eye muscle area or rib eye area; M%: Meat percentage or lean Meat; MS: Muscle score; CM: Carcass measurements; MC: Meat characteristic; KO: Kill out.
[2]In France (FRA), only meat colour for production of veal meat; in Slovenia (SVN): pH, Colour (L, a, b).

Table 3. Carcass measurement on field in abattoir.

Country	Traits [1]						
	CW	CC	Fat	FT	KPHF	CY	EMA
AUT	87 996	87 996	87 996			50 852	
AUS	*						
CHE	9 849	9 849	9 849				
DEU	381 754	381 754	381 754				
ESP	872	863	243				
FRA	30 172	30 172	30 172				
NOR	5 959						
USA	10 627		11 841	10 545	1 026		12 077

[1]CW: Carcass weight; CC: Carcass conformation; Fat: Carcass fat grade; FT: Fat thickness; CY: Carcass yield; EMA: Eye muscle area or rib eye area; KPHF: Kidney pelvic heart fat.
*Record of data exist but we haven't the figure.

Conclusion

On one side, the guidelines for beef performance recording include recommendations to record global carcass information, Warner Bratzler shear force and ultrasonic measurements. The recording of basic information of carcass weight, carcass conformation, and in the USA on ultrasonic measurement is increasing a lot on commercial carcass. This type of recording is now feasible in a larger number of countries, because of the development of a unique ID for the life of the cattle.

On the other side, there are no standardisation or guidelines for other traits related to beef quality. The major explanation is that these measures are difficult to obtain and generally expensive. So it's difficult to find their cost return in a breeding scheme, even when the industry is more and more interested by a selection on this family of traits.

The new developments in genomics offers a new way to use these specific traits related to beef quality. Several teams try to identify the genes responsible of the variability of the quality of the meat. They use large experimental designs where you can record specific traits like chemical composition, testing, etc. The next step will be to identify and to select the good alleles in the selected population using a cheap DNA test.

References

Arthur, P.F and R.M. Herd, 2006. Genetic relationships between residual feed intake and other economically important traits in beef cattle. 8th World Congress on Genetics Applied to Livestock Production, Belo Horizonte, MG, Brasil, 4 pp.

Engellandt, TH., N. Reinsch, H.J. Schild and E. Kalm, 1999. Progeny test for beef traits in German Simmental - organization, genetic parameters and breeding values. Journal of Animal Breeding and Genetics, Volume 116, 47.

Fursey, G.A.J., C.A. Miles, S.J. Page and A.V. Fisher, 1991. Speed of ultrasound in Hereford bulls and its correlation with carcass composition. Animal Production, Volume 52, 263-269.

Fouilloux M.N., G. Renand and D. Laloë, 2002. Genetic evaluation using commercial carcass data in French beef cattle. 7th World Congress on Genetics Applied to Livestock Production, Montpellier, France, n. 02-20.

Hassen, A., D.E. Wilson, V.R. Admin, G.H. Rouse and C.L. Hays, 2001. Predicting percentage of intramuscular fat using two types of real-time ultrasound equipment. Journal of Animal Science, Volume 79, 11-18.

Hohenboken, W.D., 2002. Guidelines for Uniform Beef Improvement Programs. 8th edition, Beef Improvement Federation, Manhattan, KS, USA, 165 pp.

ICAR, 2005. International agreement of Recording Practices, Section 3. ICAR Rules, standards and guidelines for meat production recording. Rome, Italy, 77-178.

Journaux, L, 2004. International beef performance recording Survey, Results for birth period 2000, Country results May 2004, Compte-rendu Institut de l'Elevage, n. 3479, 53 pp.

Journaux, L., H.J. Schild and A. Grogan, 2002. The ICAR Questionnaire for Beef Recording - Statistics and Trends. In: Crettenand, J., J. Moll, C. Mosconi and S. Wegmann (eds.), Performance recording of animals: state of the art 2002. EAAP publication 107, Wageningen Academic, Wageningen, The Netherlands, 135-138.

Journaux, L., F. Ménissier, O. Leudet, A. Hervé, M.N. Fouilloux, S. Miller, A. Havy and P. Berrechet, 2005. Plus de 30 ans d'évolution des pratiques de contrôle des performances en bovins allaitants en France. Rencontres autour des Recherches sur les Ruminants, Volume 12, 307-314.

Rehben, E., 1990. Beef recording in dairy and suckler herds. In: Gaillon, P. and Y. Chabert (eds.), Performance recording of animals: state of art 1990. EAAP publication 50, Pudoc Ed., Wageningen, The Netherlands, 1991, 111-112.

Renand, G. and A.V. Fisher, 1997. Comparison of methods for estimating carcass fat content of young Charolais bulls in performance testing station. Livestock Production Science, Volume 51, 205-213.

Renand, G., A. Malafosse, F. Ménissier, H. Leveziel, J.F. Hocquette, J. Lepetit, S. Rousset, C. Denoyelle and V. Dodelin, 2006. Genetic variability of meat quality in French Charolais, Limousin and Blonde d'Aquitaine beef cattle. 8thWorld Congress on Genetics Applied to Livestock Production, Belo Horizonte, MG, Brasil, 4 pp.

Simianer, H., H. Täubert and K. Küttner (eds.), 2001. Beef recording guidelines: a synthesis of an ICAR survey. ICAR Technical Series, Volume 6, 46 pp.

Wilson, D.E., H.U. Graser, G.H. Rouse and V. Amin, 1998. Prediction of carcass traits using live animal ultrasound. Proceeding 6th World Congress on Genetics Applied to Livestock Production, Armidale, Australia, Volume 23, 61.

Traditional and new methods to assess beef quality

Jean François Hocquette[1], *G. Renand*[1], *E. Dufour*[2], *J. Lepetit*[1] *and G.R. Nute*[3]
[1]*INRA, Unité de Recherches sur les Herbivores, UR1213, Centre de Clermont-Ferrand/Theix, F-63122 Saint-Genès-Champanelle, France*
[2]*INRA, Station de Génétique Quantitative et Appliquée, UR337, Domaine de Vilvert, F-78352 Jouy-en-Josas, France*
[2]*ENITA Clermont, Département Qualité et Economie Alimentaires, F-63370 Lempdes, France*
[3]*Sensory & Consumer Group, Division of Farm Animal Science, University of Bristol, Langford BS40 5DU University of Bristol, United Kingdom*

Abstract

Quality can be defined as 'The properties of a product that contribute to and satisfy the needs of the end-user'. Therefore, extrinsic quality attributes (including production characteristics) need to be considered in relation to how they impact on intrinsic quality attributes. The latter are focused on muscle characteristics for rearing and genetic purposes or on the product itself for sale. In muscle biology, the recent advent of genomics has the potential to identify new biological predictors of meat quality. Genomics will yield new information that will increase our knowledge of the genes affected by production factors. To assess meat quality closer to the point of sale, different indicators (pH, colour, mechanical properties, ultrasound, electrical impedance and conductivity, etc.) and technologies (image analysis, fluorescence and NIR spectroscopy, etc.) can be used. Final assessments by analytical tests with trained panels or consumer tests are methods that are chosen to evaluate meat sensory quality. Nutritional quality is assessed by the analysis of meat composition (proteins, fatty acids, etc.). In all cases, relevance with regard to the initial objectives, repeatability, accuracy, speed, on-line application and cost are essential criteria in selecting the most appropriate methods. Large scale programmes are in progress to assess and explore these criteria and the interrelationship between quality indicators. This approach may define the best combination of the indicators of quality which together with modelling approaches may produce a labelling system for meat based on quality that complies with the existing legislative framework.

Keywords: beef, sensory analysis, physical methods, biochemistry, genetics

Introduction

The great change in our Society is that the farming and agri-food sectors are faced with a general saturation of food markets in Europe and with an increasing demand by consumers for high-quality products, especially in terms of healthiness. The major current questions are thus how to define quality, and how to assess quality of animal products with the objective to satisfy the new consumer requirements. These questions are of high importance for beef (which is the subject of this review) due to the recent crises (BSE, etc.) with their effects amplified by the media. These food crises have indeed weakened the confidence of consumers in the quality of meat, although recent work in the UK (Weatherall *et al.*, 2003) showed that consumers rated 'good taste' and 'freshness' slightly higher than 'good for health'. To achieve the above objectives, the producers and retailers of animal products now need indicators for quality, as well as reliable methods to assess those indicators. Whereas indicators of beef quality have been described in a recent publication (Hocquette and Gigli, 2005), this review paper is focused on methods available for the assessment of beef quality.

The first part of this review will be, however, a summary of the ideas around the concept of quality in order to better discuss the advantages and limits of available methods for the assessment of quality. We will finally consider quality as 'the properties of a product that contribute to and satisfy the needs of the end-user'. In beef production, from the farm to the fork, the end-users may be, not only the consumers at the end, but also the abattoirs and the retailers of meat. At the consumer end, the reference methods are sensory panels and consumer panels which will be briefly described.

The second part of this review will describe some traditional and emerging physical and biochemical methods for *post-mortem* prediction of beef quality, which may be used at the abattoir or by beef retailers to ensure good quality for consumers.

The last part will deal with prediction of beef quality at the producer level on the live animal, aspects which are more and more important at the consumer end, since consumers are requesting increasing amounts of information regarding breed, sex, rearing conditions, origin and feeding systems of animals.

The concept of quality

Quality definition

Quality must now be considered as a convergence between end users' wishes and needs on one hand and the intrinsic and extrinsic quality attributes of food products on the other hand.

The first problem is the diversity of end users along the food chain. It is obvious to consider consumers as the major end-users, but consumers' wishes are not always logically consistent and vary with consumer type, incomes, country, culture, consumer's age, habits, and also with time. In addition, consumers' demands for low prices conflict with demands for high-quality products. In beef, demands for flavour tend to conflict with demands for wholesomeness and low fat. Other important end-users are retailers: they indeed play a major role by buying to producers products with specific characteristics according to their understanding of consumers' wishes and according to the market they target (for instance, low price products to eat every day or high-quality products for festive meals) including their political and economical constraints. At the end, producers have to adapt to the overall demand of consumers which have been more or less transformed by intermediaries within the different socio-economical markets.

The second problem of this definition is the increasing number of quality attributes which must be considered depending on the country culture. For instance, in the southern European countries, the concept of quality is much wider than in northern countries: it indeed includes sensorial traits, the geographical and human environment, and any link to a specific region or to any specific method of production. Besides safety (which will be not considered here), healthiness is, however, a major expectation of consumers in developed countries. Whatever the definition, most of the experts have also made a distinction between intrinsic and extrinsic quality attributes. The first refers to the product itself and includes for instance, (1) safety and health aspects, (2) sensory properties (e.g. texture and flavour) and shelf life, (3) chemical and nutritional attributes, and (4) reliability and convenience. The latter refers to traits more or less associated with the product, namely (1) production system characteristics (from the animals to processing stages including animal welfare, environmental aspects, and social considerations for instance), and (2) marketing variables (including price, brand name, distribution, origin, packaging, labelling, and traceability) (reviewed by Luning *et al.,* 2002; Grunert *et al.,* 2004; Hocquette *et al.,* 2005b). We will mainly consider some intrinsic quality attributes in this paper (not all), despite obvious strong interactions between both.

Quality model

Due to this complexity, Total Food Quality Models were proposed in an attempt to integrate a number of approaches related to quality at the consumer end (consumer perception, intention of purchase, decision-making and consumer satisfaction) including the intention to purchase a product again, which in turn depends mainly on the relationships between quality expectation and quality experience before and after purchase (Grunert *et al.,* 2004).

An example of a 'paddock to plate' quality assurance system which manages meat quality along the entire length of the meat production chain is the new grading scheme called Meat Standards Australia (MSA), which is presently being implemented for the Australian domestic beef market by Meat and Livestock Australia (MLA). MSA has identified the Critical Control Points (CCPs) from the production, pre-slaughter, processing and value adding sectors of the beef supply chain and quantified their relative importance using large-scale consumer testing. The most important CCPs have been incorporated into a model to predict palatability for individual muscles according to the cooking method since analysis of the MSA database showed that the variation in palatability explained by muscles was approximately 60 times greater than that explained by the variation between animals for the same muscle. Tenderness and tenderness-related traits are indeed highly variable among beef muscles (Rhee *et al.,* 2004). The CCPs from the production include breed, ossification (estimate of age) and implants of hormonal growth promotors. The CCPs from pre-slaughter and processing include pH/temperature window, alternative carcass suspension, electrical stimulation (both high and low voltage systems), marbling and ageing. This model confirms that a guarantee for eating quality can only be given if the links that most affect eating quality are controlled along the meat production chain (Thompson, 2002).

A strong point of the MSA is that the sensory results used are derived from untrained consumers, which means people like you and me who may buy beef every day. The current iteration of the model in June 2004 is based on a database which contains responses from 60,100 consumers. Weak points of this model are linked to the specificities of the Australian market. Indeed, farmers produce crossbred animals, and only a few pure breeds unlike in Europe, especially in France. Most of the results in the database come from steers and not young bulls or cows as in continental Europe. The breed effect is not taken into account except the *Bos Indicus* effect (vs. *Bos Taurus*). This is the reason why an initiative has been taken within a recently funded European Project entitled 'ProSafeBeef' to adapt the MSA approach to the European conditions especially in Ireland and France in collaboration with the Australian CRC on beef.

Sensory analysis

Sensory analysis is usually considered as the reference method to assess meat quality since it is a measurement of the product itself made by people who really eat meat (and hence assess its quality how real consumers would do more or less). The science of sensory analysis is relatively young when compared with the traditional sciences of Physics, chemistry, etc. The first systematic sensory test developed was the triangular test circa 1940, which was used in Scandinavian countries. Parallel development was also in progress in the USA at about this time. The first book on sensory analysis was written by Tilgner (BSI Sensory analysis of Food, part 2) in 1957 in Polish and this text was later translated into Czech, Hungarian and Russian. The second book on sensory analysis was written in Japanese, Masuyama and Miura in 1962 (BSI Sensory analysis of Food, part 3). The third textbook, which most sensory analysts will recognize, was that published by Amerine *et al.* in 1965 (BSI Sensory analysis of Food, part 4). This book was based on the lectures given at The University of California at Davies and formed part of their Food Science course programme. More recent books such as that by Meilgaard *et al.* (1991) (BSI Sensory analysis of Food, part 6) build upon previous works and include applications of difference tests that have been developed in the intervening period.

Sensory analysis is very much an interdisciplinary subject and the aforementioned works cover in some detail the basis of human sensory perception related to thresholds of determination of basic tastes, the variance in individual sensory response, plus an understanding of experimental design. Sensory analysis of meat uses people instead or in conjunction with instruments to analyse the characteristics of meat. Just in the way you would learn how to calibrate an instrument and be aware of its sensitivity, you also need to investigate the sensory ability of assessors. In industry, sensory evaluation can be used in conjunction with product developers to identify problems with a product or to optimise a number of desirable characteristics of a prototype product or in the reformulation of products caused by supply problems.

The sensory attributes of a food will provide a number of stimuli that will involve the senses of sight, smell, taste, touch and hearing. Sight is used to assess the colour, size and shape of a food; smell and taste are used to detect aroma and flavour. Whilst touch, using tactile senses and mouth feel can be used to sense the feel of a food which can include hardness, texture characteristics, viscosity, etc. Finally, hearing can be used when dealing with crunchy foods like apples or cereal bars and crisps. All of these qualities can have an influence on the quality of a food and can be assessed by sensory panels with trained assessors. Unlike sensory panels, consumer panels use untrained consumers and are concerned with preference or liking decisions.

In conclusion, sensory analysis, if used effectively, is an important stage in the development of new products or in assessing the impact on eating quality of a food by new production procedures. Whilst most managers see sensory analysis as the way to evaluate mundane questions such as 'does the new product differ from the old?', and in many ways this is the bread and butter of product testing, there are techniques available that can yield substantially more information.

Sensory panels

Companies will probably use in-house panels made up of staff members available within the company. It is important that members of the panel are screened for their basis taste acuity and colour vision and their ability to describe the characteristics of a food product. Motivation is most important and assessors must be reliable and attend all panel sessions. In this respect, it is vital that their line managers support the sensory panel. There will be occasions when it will not be possible to convene a full panel and if this becomes a regular occurrence, thought should be given to the idea of employing part-time sensory assessors. This has the advantage in that they turn up for the sessions and could prove to be a very cost effective method of obtaining sensory information. To conduct effective sensory panels, the assessors need to be free of distractions and the tests should be carried out in a special room in which there is controlled lighting and good ventilation. Assessors should be seated at separate booths and should not be able to communicate with each other during the assessments.

Presentation of samples. Samples of food should be uniform in size and at the same temperature of serving. They should be coded by a random three-digit number and presented in clean odour free containers. If more than one sample is to be assessed, then care is needed to ensure that the assessors do not receive the samples in the same order, since this will introduce a bias. Assessors are instructed to rinse their mouth out with water between each sample to remove traces of the previous sample from their mouth.

Sensory tests. These can broadly be divided into 'difference tests', 'ranking tests, 'category tests' and 'profiling tests'.

Difference tests are essentially tests where the assessor is presented with a choice situation. The paired comparison test is used to determine differences or preferences between two samples for a specified attribute e.g. tougher or more tender. These differences may be directional or non-directional. A typical question in a directional test would be 'Which sample is tougher?' Here the method requires at least 7 experts or 20 selected assessors. A typical question for a non-directional test would be 'Which of these two meat samples do you prefer?' Directional tests are one-sided (one tailed tests) whereas non-directional tests are two-sided (two-tailed tests). Triangular test is used for revealing slight differences between samples and can also be used in selecting and training of assessors. Three coded samples are presented simultaneously, two are the same and one is different. Assessors are asked to select the odd sample. The duo-trio test is an intermediate between the duo (paired) and the trio (triangular) test and is statistically less powerful than the triangle test. In this test, the assessors receive one sample marked as a reference sample and two other coded samples, and are asked which of the two samples matches the reference sample. The 'A'-'not A' test is another variation on difference testing and is used for evaluating samples having variations in appearance (when it is difficult to obtain strictly identical repeat samples), or it can be used as a perception test, to determine the sensitivity of an assessor to a stimulus. Assessors are asked to taste sample A, the sample is then removed and replaced with an even number of samples, in which there are 'A' and 'not A' samples. The numbers of correct and incorrect responses are recorded for each sample and a contingency table constructed. The chi-squared test is used to determine whether 'A' is recognised in a different way to 'not A'.

Ranking tests can be used to put samples in order of some predetermined sensory stimulus e.g. off-odour. Ranking tests have the advantage that more than two samples can be compared at the same time; whereas only two samples can be compared in difference tests, however the disadvantage is that inexact results can be obtained if the differences are very small or the samples themselves have wide variations between them. Assessors evaluate a number of samples in random order and are asked to place them in rank order based upon a specified criterion, e.g. off odour. Assessors are instructed to avoid tied rankings where possible. Results are then collated and the rank sums for individual samples calculated. The Friedman test, F (analysis of variance by ranks) gives the maximum opportunity for showing recognition by assessors of differences among samples (Friedman, 1937). If F is equal or greater than the critical value corresponding to the number of assessors, the number of samples and the selected level of significance, it can be concluded that there is an overall difference between the samples.

Category scales are often used to rate a number of different stimuli, e.g. texture, flavour, juiciness etc. They have the advantage that they are easy to use and can cover a number of attributes, as well as up to six samples in a session. For the test to be successful, it is necessary to have good experimental design. The results can then be analysed, for example, using analysis of variance techniques to establish the differences between and within the samples. The use of these types of category scales are useful for determining the gross differences in eating quality and are frequently used in animal production experiments that require sensory data. However, this approach may not yield sufficient information if there is a need to identify the individual characteristics of a food, in which case it is necessary to derive a profile of the food.

Sensory profile methods collect various ways of carrying out descriptive analysis of foods, and there are two basic methods of profiling. In the fixed choice method, the assessors each develop a vocabulary to describe the food under test. One of the assessors acts as panel leader and their task is to discuss the words generated by the others assessors so that a consensus can be agreed. The agreed set of descriptors is then used to describe the food. The free-choice method involves the assessors being given the likely range of samples during a number of training sessions to derive a profile that is unique to them. The assessors then use this profile to describe the foods in the experiment. In both of these methods, it is important that the experiments are well designed statistically since if they are not it is unlikely that meaningful results will be obtained.

Consumer panelling

Unlike sensory panels where the taste acuity of assessors is known and results are derived essentially from differences or intensities in food characteristics, consumer panels are concerned with preference or liking decisions. There are a number of ways in which data can be obtained, but in all cases greater numbers of participants are required in consumer panelling for the data to be reliable, typically in excess of 100 people. There are three common methods of obtaining consumer data: 'home tests', where individuals receive and prepare samples in their homes; 'hall tests', where consumers are invited to assess foods under similar conditions to a sensory panel; and 'in house tests', where employees of the company are invited to take part in consumer assessments either in the workplace or at home. Each of these tests have advantages and disadvantages, and are often affected by outside factors, such as cost, which may have a significant bearing on the method used. Hall tests are now becoming increasingly used because they enable the researcher to obtain comparative data from a consumer for different examples of the product under test. They have the advantage that the researcher maintains control of the product and its preparation, and the consumer is not influenced by other members of their family, which is often the case in home tests.

Traditional and emerging physical methods for *post-mortem* prediction of beef quality

The quality of meat comes from some of its physical and bio-chemical properties. Therefore physico-chemical methods have long been used for meat quality prediction (Chrystall, 1994). The reader is invited to read previous papers to get a more complete view of such methods (Monin, 1998; Mullen and Troy, 2005). These methods should be accurate, not so expensive, rapid, with the capability to be fully automated, if possible non-invasive and should bring a benefit compared to other available methods by being related to the desired quality traits. In practice, although many such methods have been proposed, very few meet the requirements of industry. It was hypothesized in 1998 that ultrasonic measurements, image processing and NIR spectroscopy would have be the most promising methods for large-scale meat quality evaluation. Nowadays, the industry can develop mechanical, electrical and optical methods. Some of them will be described here in more details. Image processing which may have ability to assess basic traits such as colour and marbling will not be described here. Similarly, pH and the methods to measure it will not be treated as they have been described in previous review papers (Monin, 1998; Mullen and Troy, 2005).

Mechanical methods

Meat tenderness comes from the mechanical properties of muscle fibres and connective tissues as well as from interactions between them. Numerous mechanical methods have been devised to give an estimation of meat tenderness. They are based on shear, tensile, compressive or torsional strain (for a review see Bouton *et al.,* 1975, and Lepetit and Culioli, 1994). Although each method has its specificity in its ability to detect the influence of some factors on meat tenderness, these methods show significant correlation between each others (Lu *et al.,* 1998). Furthermore, myofibrillar resistance on day 2 is predictive of that on day 8, i.e. when the meat has reached approximately 92% of the tenderization (Lepetit and Hamel, 1998). In contrast, the correlations are very low between mechanical properties of raw and cooked meat even with the same method (Hocquette *et al.*, 2001) because the muscle structure (especially that of connective tissue) is changed by cooking (McCormick, 1999).

There is not a best mechanical method for meat tenderness evaluation. Nevertheless, the Warner Bratzler shear device which is an old method (Warner, 1928) is used as a reference because for long time it has been widely available commercially. However, it is an empirical method, the mechanical significance of the parameter measured is far from clear and it has been largely criticised (Voisey, 1976). Although these objective measurements (such as shear force) have the advantage of being relatively cheap, they are rather simplistic one dimensional measures of a complex set of interactions which occur when cooked meat is chewed and masticated in the mouth. Furthermore, studies in France showed that shear force may explain only up to 48% of total variability in tenderness (assessed by sensory panels) and this proportion depends on the breed and the production system (Brouard *et al.*, 2001).

Warner Bratzler shear force of rapidly cooked beef can be used in industry to classify meat. A measure of shear force at 3 days post-mortem predicts the tenderness after 14 days with 94% accuracy (Wheeler *et al.*, 1998). Taking a sample of meat for the measurement has a cost which rises the price of meat, but it has been demonstrated that consumers will pay more for meat if they have a guarantee that it is tender (Boleman *et al.*, 1997).

On the other hand, a number of invasive probes have been developed for the prediction of meat tenderness from measurement on the carcass, the first one being the Armour Tenderometer (Carpenter *et al.*, 1972; Harris *et al.*, 1992). A modified version of the Armour tenderometer (Timm *et al.*, 2003) with six needles gave a significant correlation with a sensory panel (r=-0.74). George *et al.* (1997) and George (1998) have tested the Swatland's Probe and the TenderTec to predict tenderness of beef from measurements on carcasses. Correlations between Tender·Tec readings (r=0.19) and Swatland's Probe readings (r=0.17) with shear force values for steaks from those carcasses revealed that neither of those instruments was capable of predicting meat tenderness. They concluded that, as has been reported for Tender Tec's predecessors (STE Press, Nip Tenderometer, Armour Tenderometer, MIRINZ/MRC Tenderometer) tenderness prediction capabilities from measurements on carcasses appear limited because of the inability of these measures to characterize the changes which occur in muscles during the cooking process (George *et al.*, 1997).

Electrical methods

The electrical impedance is the property of a material that limits the propagation of the current. When this property does not depend on frequency it is a resistance but in biological tissues it depends on frequency and has resistance and capacitance components. The impedance comes from the capacitance of the myofibre's membrane and from the resistances of intra and extra cellular liquids. The determinant role of plasma membrane in meat electrical properties has been frequently mentioned in the literature, but the membrane of sarcoplasmic reticulum may also act as an isolator (Swatland, 1987). Therefore impedance measurements are able to detect any changes or damage in membranes and modifications in intracellular space which occur in processes e.g. rigor onset, ageing, freezing or cooking.

During the onset of rigor, there is an increase in membrane permeability and in extra cellular space which lead to a decrease in impedance. These changes are linked to the rate of pH fall and to ultimate pH. Several electrical devices have been developed to predict pH allowing detection of defaults in pork or turkey such as pale soft exudative (PSE) or dark firm dry (DFD) meat. Different variables have been considered such as the electrical conductivity, the capacitance or the dielectric loss factor (which is the ratio of the conductivity to the dielectric constant). Although some systems are used at the slaughter line, their validity has been criticised as it is not always possible to differentiate between normal and PSE meat (Schmitten *et al.*, 1984; Fortin and Raymond, 1988) and the correlations obtained with quality vary greatly (Forrest *et al.*, 2000; Garrido *et al.*, 1994; Jaud *et al.*, 1992).

Freezing affects meat impedance because during freezing ice crystals damage the membranes. Consequently, after freezing the electrical impedance is low and depends only on intra and extra cellular liquids. Therefore, the measure of impedance has been proposed as a means for detecting meat that has been frozen (Salé, 1972).

A number of studies have investigated the use of electrical properties to evaluate the amount of fat in carcass or in meat. Fat is an extremely good insulator and therefore influences impedance. Measurements of carcass composition with electrical measurements have led to very good predictions. However, the predictive equations require the use of morphological parameters. Forrest *et al.* (1989) obtained an r^2=0.89 for the prediction of pork carcass lean with electrical measurement associated with carcass length and temperature. Swantek *et al.* (1992) showed that 82 to 84% of the variation of fat free mass of pork carcasses was explained by electrical measurements associated with carcass weight. Similar results are obtained with measurements on live animals using live weight. Jenkins *et al.* (1995) predict 94% of the variation in fat free soft tissue of steers and heifers with impedance, carcass weight and fat depth at the 12th rib.

After a drop at rigor, electrical impedance continues to decrease during ageing but much slowly (Salé, 1976; Pliquett *et al.*, 1995; Lepetit *et al.*, 2002). No clear understanding of *post-rigor* variations in impedance has been given. The ratio of impedance at 2 frequencies (Z_{1kHz}/Z_{100kHz}) has been proposed as an ageing index since it is independent of fat level. Although the decrease of this parameter during ageing is linearly related to the decrease in muscle fibre resistance, its dependency with muscle fibre resistance depends on muscle type (Lepetit *et al.*, 2000). On the other hand, the relationship between electrical anisotropy of muscle and muscle fibre resistance is independent on muscle type. The electrical anisotropy can be defined as the ratio of the impedance measured in the direction perpendicular to muscle fibres to the impedance measured in the direction of muscle fibres. This ratio equals about 2 close to rigor and tends to 1 during ageing and has been proposed as a non destructive measure of the state of ageing of meat (Lepetit *et al.*, 2000; Patent N° 0453206).

Optical methods: fluorescence and NIR spectroscopies

The use of spectroscopy in food science has increased tremendously in the last couple of decades as it has appeared that detection and estimation of a number of food constituents and properties may be achieved by measuring the amounts of the radiation that are either absorbed or emitted at different wavelengths. Absorption spectroscopy is now widely used in food analysis. Emission spectroscopy has increased much in importance in the last decade, and is presently in wide use in the estimation of certain elements in food products.

In recent years, a range of equipment for rapid analysis of meat have been developed, based on the principles of x-rays, fluorescence and NIR spectroscopies, electrical impedance and ultrasonics (Hildrum *et al.*, 2006). Near-infrared (NIR) spectroscopy is one example of a method that is successfully used to quantify composition and as a tool for prediction or classification of product quality (Isaksson *et al.*, 2001). Recently, fluorescence has also proven potential for rapid and non-destructive analysis of foods (Dufour and Frencia, 2001).

Fluorescence spectroscopy offers several inherent advantages. First, it is 100-1000 times more sensitive than other spectrophotometric techniques. Second, fluorescent compounds are extremely sensitive to their environment. For example, tryptophan residues that are buried in the hydrophobic interior of a protein have different fluorescent properties than residues that are on a hydrophilic surface. This environmental sensitivity enables characterisation of conformational changes such as those attributable to the thermal, solvent or surface denaturation of proteins (Lakowicz, 1983), as well as the interactions of proteins with other food components. Third, most fluorescence methods are relatively rapid and a spectrum is recorded in less than 1 second with a CCD detector.

Fluorophores can be broadly divided into two main classes: intrinsic and extrinsic. Intrinsic fluorophores are those which occur naturally in the product sample. In meat these include the aromatic amino-acids tryptophan, tyrosine and phenylalanine, structural proteins such as elastin and collagen, the enzymes and coenzymes NADH, FAD and NADPH, the vitamins A, K and D, derivatives of pyridoxal, porphyrins, phospholipids and the lipid pigments lipofuscin and ceroids (Ramanujam, 2000). Riboflavin (also named vitamin B2) is another prominent fluorophore, that occurs in meat and milk products (Karoui *et al.,* 2005).

Recently, front face fluorescence spectroscopy in combination with multivariate statistical methods has been more commonly used for studying quality parameters of 'native' samples of milk, cheese, meat and meat products. Although many of these studies indicate interesting potentials for solid sample measurements, the implemented applications are, however, rather rare. Munck's review (1989) shows the versatility of fluorescence techniques for quality assessment of cereals, meat and fish. Within the field of fluorescence spectroscopy in meat science, Swatland is one of the main contributors. Through numerous articles on fluorescence properties related to meat quality parameters, connective tissue in particular, he pinpointed the potential of direct measurements on meat (Swatland, 1987, 1991, 1993, 1996). It has long been known that connective and adipose tissues are autofluorescent. It is well known that different types of collagen crosslinks such as hydroxylysyl pyridoline, lysyl pyridinoline and pentosidine are contributors. Collagen exists in several different genetic forms, four of which have been found to be present in muscle; types I, III, IV and V. Types I, III, and IV have similar fluorescent properties for excitation in the region 330-380 nm, while type V differs from the others. Another powerful fluorophore in meat is elastin, which exhibits fluorescence quite similar to collagen types I, III IV (Wold *et al.,* 2000; Egelandsdal *et al.,* 2005). Adipose tissues contain a connective tissue network, but also other fluorescent components more specific for fat, such as the age-related pigments lipofuscin and ceroid (Yin, 1996). The fat soluble vitamins A, D and K exhibit fluorescence in the region 387-480 nm when excited in the 308-340 nm region (Dufour and Riaublanc, 1997; Ramanujam, 2000; Skjervold *et al.,* 2003).

Oxidation of unsaturated lipids in meat involves many complex redox reactions, and a variety of lipid oxidation products are formed. Some of the oxidation products are unstable and will react with other compounds in the matrix, leading in several cases to fluorescent reaction products. These compounds are formed from reactions of oxidising fatty acids or lipid oxidation breakdown products (hydroperoxides and aldehydes) with compounds containing primary amino groups (proteins, amino acids, DNA) (Kikugawa and Beppu, 1987). Fluorescence is by now recognised as a sensitive method for determining the level of lipid oxidation in complex foods such as fish and meat (Melton, 1983). Recently it has been reported that good estimates of the degree of rancidity in ground poultry meat as well as on more complex meat products, can be obtained by fluorescence measurements directly on the product (Wold and Mielnik, 2000; Wold *et al.,* 2002).

The tryptophan fluorescence spectra from meat have also been evaluated for tenderness measurement and for discrimination between different muscle types (Dufour and Frencia, 2001). Lebecque *et al.,* (2003) showed that the texture attributes of meat may be derived from the fluorescence spectra of protein tryptophans. Allais *et al.* (2004) also showed that similar results could be obtained on meat emulsions and frankfurters.

Near infrared (NIR) spectroscopy is one of the techniques that utilize the vibrational energy transitions of molecules. In the NIR region (780 to 2500 nm) of the electromagnetic spectrum, the observed molecular vibrations are overtones and combinations of the fundamental vibrations of the mid-infrared spectral region. NIR applications regarding meat composition has been reported for a long time (Berge, 2006; Downey and Hildrum, 2004). The prediction errors for fat, water and protein have typically been in the range 0.3-0.7% for off-line applications. During the last decade a number of dedicated NIR instruments have appeared on the market (Isaksson *et al.,* 1996; Schwarze, 1997). On-line NIR prediction of fat, water and protein in large industrial scale batches of beef and pork was further studied by Togersen *et al.* (1999). Near-infrared spectroscopy also has the ability to reveal changes in the state of water and hydrogen bond interactions in food. Such changes occur in beef during ageing, which makes NIR an interesting candidate for beef tenderness assessment (Berge, 2006; Downey and Hildrum, 2004). However, careful consideration on how to obtain representative sampling is critical when developing applications based on NIR spectroscopy. Most foodstuffs are highly heterogeneous with regard to distribution of major constituents, such as collagen, fat and protein. This is a potential problem when one wants to obtain representative NIR measurements to get the best possible estimate of the chemical composition.

In conclusion of this part, it is clear that fluorescence and NIR spectroscopic techniques have great potential for applications in the meat sector. A large part of the studies, which are reported above, are based on models with a limited number of independent samples. However, the success of the presented research strongly supports the contention that these applications can be successfully developed for rapid measurements of meat quality.

Assessment of beef quality at the production level on the live animal

The critical control points underpinning palatability are directly dependant on key biochemical parameters including on one hand, muscle characteristics of live animals and, on the other hand, *post-mortem* muscle biochemistry which is associated with ageing (Maltin *et al.,* 2003). The former include glycogen metabolism (which affects *post-mortem* pH and hence also ageing), the initial levels of the various proteolytic enzymes involved in protein degradation during ageing in association with muscle fibre characteristics, intramuscular adipocyte number and size (marbling), connective tissue chemistry (basal toughness), and pigment content which contributes to colour.

The traditional view is that these biochemical characteristics explained no more than one-quarter to one third of beef tenderness or flavour (Renand *et al.,* 2001). This argues that more research is needed to identify new muscle properties which may influence meat quality traits in order to control a greater part of its variability. Researchers are currently taking advantage of the recent development of functional genomics to achieve this objective. This is important for producers and farmers which are looking for rearing factors which regulate muscle characteristics and hence the potential of the animals to produce meat of high and constant quality.

Other end-users of the muscle knowledge are geneticists and breeders who are looking for muscle characteristics that are both associated with quality and heritability which can be passed through the generations. Both geneticists and breeders have already taken advantages of the huge progress in structural and functional genomics to achieve this goal (Hocquette, 2005a).

Biochemical characteristics of the muscle tissue

Among the muscle characteristics linked to the metabolism of live animals, the pH of meat (which depends on glycogen metabolism before slaughtering) and intramuscular fat content are the major factors associated with beef quality traits. For instance, studies with young bulls in France have shown that pH at 3 hours *post-mortem* and intramuscular fat content explain 52% and 56% of the variation in tenderness and flavour respectively (Dransfield *et al.,* 2003). Furthermore, any change in growth rate (Cassar-Malek *et al.,* 2004) and feeding conditions (Listrat *et al.,* 2001) were shown to affect primarily muscle metabolic activity.

All these muscle characteristics can be determined more or less accurately by laboratory methods, most of which are often time consuming and costly. A great deal of research is indeed being conducted to improve available methods for the assessment of the known indicators of quality and / or to search for new indicators of quality. For instance, in the connective tissue area, after the assessment of total and insoluble collagen amount in muscles, some other methods have been described to determine the amounts of type I, type III (major collagen isforms) (Listrat and Hocquette, 2004) and type XII and XIV (minor collagen isoforms) (Listrat *et al.,* 2006) in muscle tissues. However, the biology of the connective tissue is so complex that many of its characteristics are potentially important for meat quality (Purslow, 2005). The same reasoning can be made for muscle fibre type (Picard *et al.,* 2002) and muscle metabolism (Hocquette *et al.,* 1998). For the determination of muscle fibre type based on their contractile activity, different methods of histochemistry, immunohistochemistry and electrophoresis are hence available (Picard *et al.,* 2002). For muscle metabolism, different enzymatic methods are also currently used (Hocquette *et al.,* 1998).

The problem in muscle biology is thus firstly to demonstrate the importance of the assessed criteria for quality and secondly to convert the available methods into routinely used tests. The two questions are in fact linked together. In practice, biologists carefully choose the best technical options to get a better understanding of the muscle biology and to specifically assess the quality markers of interest. In other words, experts in muscle biology are usually efficient in discovering new methods. Unfortunately, no such close attention is usually given to the calculation of variability which may impair the results. We thus recommend calculating the number of identical samples per group to be analysed as well as the number of repetitive measurements per sample to be able to detect any true biological difference between groups depending on the biological and technical variability of the studied variable (Hocquette, 2004). For instance, a recent study was performed in France with about 1200 samples of *Longissimus thoracis* muscle from different young bulls (programme QUALVIGENE funded by AGENAE and coordinated by UNCEIA; Table 1). Some variables were measured in duplicates (intramuscular fat content, muscle fibre area), others in triplicate (collagen contents), other five- (colour) or ten-fold (shear force) per sample. These differences may be explained by the low accuracy of some methods (for instance, collagen content, especially in muscles poor in connective tissue; Listrat and Hocquette, 2004) or by the heterogeneity of the biological material (for instance, for muscle fibre area). By comparison, tenderness, juiciness and flavour (determined by 12 sensory assessors) show a much greater variance among assessors than among animals, the unexplained variance being also high. These results are in favour of using direct biochemical measurements related to the muscle tissue itself rather than sensory analysis providing a strong association with the final quality. However, it is important to realise that a trained sensory assessor when considering tenderness is only concerned with the global appreciation of the relative toughness/tenderness of the sample. If it is a requirement to describe the characteristics of texture, then a full descriptive profile is necessary. The correlations with individual components of muscle will then substantially improve.

As stated before, the variation in palatability between muscles is greater than that between animals for the same muscle (Stolowski *et al.,* 2006). This is mainly explained by huge differences in muscle characteristics as previously demonstrated. While the assessment of palatability attributes of the major beef muscles have been published (Jeremiah *et al.,* 2003), the biochemical characteristics of 39 different muscles were also described (Von Seggern *et al.,* 2005). Scientists now have access to databases in which all the current available knowledge about average muscle properties (http://bovine.

Table 1. Explanation of the variance for several measurements of muscle and meat quality with about 1200 samples of Longissimus thoracis muscle from different young bulls (programme QUALVIGENE funded by AGENAE and coordinated by UNCEIA, France).

Meat quality parameters	Animals (n)	Measurements (n)	Part of the variance (%)	
			explained by animal effect	not explained by animal effect
Intramuscular fat content (%)	1233	2467	93	7
Collagen content (%)	1233	3697	61	39
Muscle fibre area (μ^2)	1236	2472	81	19
Colour: L*	1249	6287	85	15
Colour: a*	1249	6287	87	13
Colour: b*	1249	6287	88	12
Shear force (kg)	1197	11950	38	62

Breed	Measurements (n)	Part of the variance (%) in								
		Tenderness explained by			Juiciness explained by			Flavour explained by		
		animal	sensory assessor	residual	animal	sensory assessor	residual	animal	sensory assessor	residual
Charolais	4878	19	32	49	4	57	39	2	47	51
Limousins	4991	8	25	67	8	35	58	4	23	73
Blond d'Aquitaine	4343	16	17	68	12	17	71	3	18	79

unl.edu/eng/). It was clearly shown that breed type influences muscle characteristics of individual muscles, but this role has been shown in one study to be limited to the rate of *post-mortem* aging and calpastatin activity (Stolowski *et al.*, 2006), except for specific genotypes characterised by lower muscle collagen content (i.e. double-muscled cattle). Similarly, data bases for the nutritional value of beef based on muscle composition in fatty acids, minerals, etc would be very useful to meet the increasing requirement for healthiness by consumers. The muscle profiling data will allow for more informed decisions to be made in the selection of individual muscles for the production of value-added values, for the choice of the most effective *post-mortem* intervention techniques or for the choice of the most adapted cooking methods in order to increase the general consumer acceptance of beef.

Functional genomics applied at the muscle tissue

It has been known for many years that gene expression profiling at the mRNA and protein level will provide a better understanding of the mechanisms that underline the development of muscle biochemical characteristics related to quality depending on genetic and environmental factors. This is simply due to the fact that muscle biochemical characteristics depend on gene expression in the biological world.

The recent advent of high-throughput techniques in molecular biology has given to scientists the possibility to analyze thousands of genes or proteins simultaneously (Mullen *et al.*, 2006). Similar recent techniques were also developed for the detection of bacterial contaminations (for beef safety) or species and individual identification (for traceability) but these aspects are outside the scope of this paper despite their great interest for the beef industry. DNA arrays are nowadays the most popular approaches to transcriptomic studies, which aim to quantify gene expression at transcript (or mRNA) levels. Two-dimensional electrophoresis followed by mass spectrometry is the most popular approach to proteomic studies, which aim to quantify gene expression at the protein level. Many papers have described the advantages and limits of those techniques and also their application to meat quality. Only a few major conclusions will be recalled here.

A strong point of all these approaches is that they give access to a great number of information (expression levels of thousands of genes and proteins). This has two major advantages. Firstly, providing good statistical correlation studies, the comparison of tough and tender samples (or of tasty and not tasty samples) will evidence differentially expressed genes or proteins which might explain a greater part of the variability in tenderness (or flavour) than traditional biochemical characteristics. This may be the case, since a recent experiment conducted in France has indeed evidenced a gene, which has been patented, whose expression level explains almost twice the variability in tenderness than any other biochemical measurement. Markers of tenderness, which are different between breeds, have also been revealed by proteomic studies (Bouley *et al.,* 2004). Another study has revealed a biological indicator of marbling (Barnola *et al.,* 2005).

A weak point of these approaches is their actual complexity since the technology is new. One of the most important stages of any genomic experiment is the initial planning. It involves factors such as the sample source, the number of replicates, the type of platform and the experimental design. In transcriptomics, the first DNA arrays were generally produced 'in house' by academic groups. They were cheap but required large amounts of optimization time and expensive equipment. Commercial platforms are more expensive, but more accurate, but often not as flexible as 'in house' arrays. Reproducibility for most platforms within any lab is generally good, but poorer between platforms and across labs. But, reproducibility is improving by increasing standardized procedures (Elvidge, 2006), which is nowadays the great challenge of the biggest companies. Advantages and limits of genomic approaches have been extensively reviewed by many authors (Hocquette *et al.,* 2005a, Mullen *et al.,* 2006), as well as the current state of knowledge and important lessons derived from bovine functional genomic work (Lehnert *et al.,* 2006). Similarly, proteomics is now helping in the discovery of protein markers of the various quality traits (Bendixen, 2005). Among the difficulties of both transcriptomics and proteomics are surely statistical analyses because high-throughput data raise unique issues which are a great matter of debate (Shoemaker, 2006).

Another weak point (but also a strong point) of genomic studies is that they give access to a defined list of genes or proteins associated with the quality trait of interest. For the researcher, the difficulty is now to extract meaningful biologic significance from the results. In addition, genes and proteins do not function independently, but interact together that give rise to cellular functions, and hence to muscle characteristics and the resulting beef quality. Gene and protein expression profiles should be thus studied in association with muscle biochemistry and beef quality studies using appropriate bioinformatic tools. This is System Biology, the science of integrating data from many levels of biology and many technologies in order to understand complex systems (Hocquette, 2005), beef quality being one of them.

Genetic determinants of beef quality

Genetic improvement is quite effective because it is permanent and cumulative: indeed improvements made in one generation are passed to the next ones. The variation in quality arising from genetic variability is far from small (0.20 - 0.25 on average in tenderness, shear force, water holding capacity, reviewed by Renand, 1988; Marshall, 1999 and Burrow *et al.,* 2001). Because quality data based on the previous described methods are not recorded in the slaughterhouse due to their excessive cost, breeding organizations cannot include direct meat quality traits in their genetic improvement programs. But, in a competitive commercial environment, the benefits to be gained from selection for slight improvements in quality of beef are significant enough to have resulted in a great of research in this area. In addition, emphasis is nowadays put on genetics and genomics due to the sequencing of genomes from different species, the bovine one being available in the very short-term future. This will have direct applications in genetic improvement and traceability with the help of DNA-based techniques.

The difficulties in working with classical selection methods enhance the role for working directly on the genetic variability using polymorphisms in genes, mainly Single Nucleotide Polymorphisms (SNP) (Renand *et al.,* 2002). In this way, the selection of sires for improving meat quality could be done on molecular information alone providing a single blood sample for DNA extraction (which is much easier to get compared to a biopsy sample for biochemical or gene expression studies). This implies as a prerequisite to have adequate information about DNA markers that are the causal mutations themselves, or, at least, SNPs within functional genes that are in close linkage disequilibrium with the causal mutations in the genes responsible for the observed genetic variability. The availability of exploitable DNA markers would allow the improvement of beef breeds through selection by increasing the frequency of the favourable alleles. In addition DNA markers will be advantageously used for matching the beef production and marketing systems most suited to the genetic potential of beef cattle according to their molecular information. Animals with favourable alleles could receive a certification of quality of their meat in high quality marketing.

Selection of animals using DNA markers requires previous research works for localizing genome regions associated with QTL (Quantitative Trait Locus) participating to the genetic variability of traits. QTL have been found, but results are difficult to synthesize. Communication does often lack details and results are generally specific of the resource families used for the primo detection. QTL can be exploited in Marker Assisted Selection (MAS) programs as long as the QTL and the linked markers are actually segregating in the production populations. However, as MAS requires the phase between the markers and QTL being established within each family, there is a need for further recording of phenotypes, i.e. meat quality measures. That recording is undoubtedly a strong limiting factor for MAS implementations.

Availability of DNA markers in close disequilibrium with the functional mutation in the beef cattle populations will make selection much easier, since no or few testing of relatives will be needed. The next research is therefore on the fine mapping of the QTL with a denser use of new markers. Apart from 3 or 4 papers, very few results are published because detection of close disequilibrium markers is commercially strategic. Actually, most of research teams are focused on searching for positional candidate genes preferentially, using the comparative genome cartography, and then detecting SNPs markers within or close to these genes which can then be used as genetic tests. Research of candidate genes was effectively conducted by different research teams for detecting and patenting SNP markers potentially useful for meat quality improvement. Published or patented DNA markers are already available for genetic tests, marketed by genomics companies, for improving beef meat quality. The list of these companies and tests can be found in *http://animalscience.ucdavis.edu/animalbiotech/biotechnology/mas/*.

The actual utility of these potential markers for improving beef cattle populations requires validation studies previously to any implementation:

a. validating if QTLs are actually segregating in the cattle population of interest;
b. estimating allele frequencies of the markers in the population of interest;
c. evaluating if the linkage disequilibrium between the markers and the performance or quality traits are close enough to be used.

Large scale Research and Development programs are obviously necessary for providing complete information on methods that can be efficiently used for the genetic improvement of meat quality: estimation of genetic variability, identification of outstanding sires and validation of DNA markers linked to QTL or in candidate genes. Such programs were implemented in Australia (Cooperative Research Center (CRC) for the Cattle and Beef Industry, 1993-2000; Bindon, 2001) and United States (Carcass Merit Project (CMP), 1998-2003; Moser *et al.,* 2004) in their more important beef cattle populations and their predominant production systems. In Australia, seven breeds were included in the CRC program, 4 British and 3 tropically adapted breeds and 7415 progeny sired by 388 sires were recorded for meat quality traits, which genetic parameters were published by Johnston *et al.,* 2003. In USA, more than 8,200 progeny of over 300 sires representing 14 breeds were recorded for carcass and meat shear force and first genetic parameters were communicated (Thallman *et al.,* 2003). The DNA and phenotype data of 2,615 progeny of 70 sires was also used for validating QTL effects. Seven among the 11 QTL detected in the Texas University QTL detection program were actually significant in these breeds (Thallman *et al.,* 2003). In France, a program is in progress with approximately 3,550 young bulls from 130 sires placed in progeny testing between 2003 and 2006 on the basis of an average of 25 progeny tested per sire.

We can see from the above examples that the efficient identification of the genes and their associated markers that influence QTL requires large sample sizes. It also needs analyses of a large number of polymorphisms in and around genes suspected to be involved in the DNA region of interest. Repeated testing in independent samples from different animal populations is also important. The choice of the quality trait to be improved is critical to the successful transition from QTL to diagnostic tests based on SNP discovery. This trait has to be chosen firstly according to its economic importance. Secondly, no alternative methods to improve this trait, cheaper to implement than a DNA test, should exist for the DNA test being worth commercializing (Barendse, 2005).

Conclusion

Due to the increasing complexity of the quality concept, more and more methods are required to assess the different quality traits of beef at the producer, the retailer and the consumer levels. Those methods differ in their objectives according to the targeted end-users, in their accuracy and in their potential to be automated. Considerable efforts have been made to improve and standardise methods of determining beef quality, but some additional efforts have still to be done for most of them. In addition, there is still a largely unsatisfied demand for low-cost and rapid routine tests for the determination of beef quality. The best methods to assess quality indicators should be non-invasive, not expensive, automated, accurate and with an assured benefit by being directly related to the trait of interest, but this is much less easy to assess routinely. In all cases, factors such as experimental design, sample numbers and conditions, sampling methodology, instrument type, data analysis and statistical issues are critical to assess the interest of any method. In the last decades, three major areas have being evolving rapidly bringing new hopes for the future.

The first area deals with methods for the *post-mortem* prediction of quality at the industrial level. Many biochemical parameters and physical measurements have already demonstrated powerful predictive ability. Emerging technologies such as electrical and optical methods have been recently developed (and some of them patented). They aim at determining the structure of meat based on its physical properties. But, in many cases, the results are difficult to be biologically interpretable. The application of any of these techniques to on-line situations will require large scale industrial trials.

A new science called genomics is likely to impact our knowledge of ruminant physiology and genetics. Genomics aims to bring a better understanding of how biological traits are determined from genes. Genomics is changing our scientific paradigm, because the global expression of genes in cells and tissues will generate new biological hypotheses. We will thus move from hypothesis-driven research (where scientists test the relevance of biological hypotheses) to hypothesis-generation research (in which new biological hypotheses will appear from the gene expression profiles). In addition, genomics has a great potential for the discovery of new DNA markers which might be used as low cost and easy-to-use diagnostic tests for the improvements of livestock breeding. This area is commercially strategic and many DNA genotyping methods for specific genes have been already patented.

The last area of interest is the advent of modelling systems which aim at predicting beef quality from information from the animal birth to cooking through animal growth and slaughtering, and also through beef processing during and after ageing. The development of such models requires very large sample sizes, the recording of all the requested data from the farm to the fork and powerful statistical analysis for the identification of critical control points of quality along the food chain. The choice of the trait(s) to be improved (for instance tenderness which is among the most important ones) is also very important as well as the method to assess it (asking untrained consumers for instance). Adaptation of such systems to the specificities of markets is also a key issue.

References

Allais, I., C. Viaud, A. Pierre and E. Dufour, 2004. A rapid method based on front-face fluorescence spectroscopy for the monitoring of the texture of meat emulsions and frankfurters. Meat Science, Volume 67, 219-229.

Amerine, M. A., R.M. Pangborn and E.B. Roessler, 1965. Principles of Sensory Evaluation of Food. Academic Press, New York, USA.

Barendse, W., 2005. The transition from quantitative trait loci to diagnostic test in cattle and other livestock. Australian Journal of Experimental Agriculture, Volume 45, 831-836.

Barnola, I., J.F. Hocquette, I. Cassar-Malek, C. Jurie, G. Gentès, J.F. Cabaraux, C. Cuvelier, L. Istasse and I. Dufrasne, 2005. Adipocyte fatty acid-binding protein expression and mitochondrial activity as indicators of intramuscular fat content in young bulls. In: Hocquette, J.F. and S. Gigli (eds.), Indicators of milk and beef quality. EAAP Publication 112, Wageningen Academic Publishers, The Netherlands, 419-424.

Bendixen, E, 2005. The use of proteomics in meat science. Meat Science, Volume 71, 138-149.

Berge, Ph., 2006. Application de la spectroscopie proche infrarouge à l'analyse des viandes et des produits carnés. In: Bertrand, D. and É. Dufour (eds.), La spectroscopie infrarouge et ses applications analytiques. Lavoisier, Tec & Doc, 2ème édition, 627-646.

Bindon, B.M., 2001. Genesis of the Cooperative Research Centre for the Cattle and Beef Industry. Australian Journal of Experimental Agriculture, Volume 7, 843-853.

Boleman, S.J., S.L. Boleman, R.K. Miller, J.F. Taylor, H.R. Cross, T.L. Wheeler, M. Koohmaraie, S.D. Shackelford, M.F. Miller, R.L. West, D.D. Johnson and J.W. Savell, 1997. Consumer evaluation of beef of known categories of tenderness. Journal of Animal Science, Volume 75, 1521-1524.

Bouley, J., B. Meunier, J. Culioli and B. Picard, 2004. Analyse protéomique du muscle de Bovin appliquée à la recherche de marqueurs de la tendreté de la viande. Rencontres autour des Recherches sur les Ruminants, Volume 11, 87.

Bouton, P.E., P.V. Harris and W.R Shortose, 1975. Possible Relationships between Shear, Tensile and Adhesion Properties of Meat and Meat Structure. Journal of Texture Studies, Volume 6, 297-314.

Brouard, S, G. Renand and F. Turin, 2001. Relations entre caractéristiques musculaires et tendreté dans le cas du muscle *longissimus lumborum* de jeunes bovins de races rustiques. Rencontres autour des Recherches sur les Ruminants, Volume 8, 49-52.

BSI Sensory analysis of food, 1982. Part 2. Paired comparison test. BS 5929, British Standards Institution, London, UK.

BSI Sensory analysis of food, 1984. Part 3. Triangle test. BS5929, British Standards Institution, London, UK.

BSI Sensory analysis of food, 1985. Part 4. Flavour profile methods. BS5929, British Standards Institution, London, UK.

BSI Sensory analysis of food, 1989. Part 6. Ranking. BS 5929, British Standards Institution, London, UK.

Burrow, H.M., S.S. Moore, D.J. Johnston, W. Barendse and B.M. Bindon, 2001. Quantitative and molecular genetic influences on properties of beef. Australian Journal of Experimental Agriculture, Volume 41, 893-919.

Carpenter, Z.L., C. G. Smith and O.D. Butler, 1972. Assessment of Beef Tenderness with the Armour Tenderometer. Journal of Food Science, Volume 37, 126-129.

Cassar-Malek, I., J.F. Hocquette, C. Jurie, A. Listrat, R. Jailler, D. Bauchart, Y. Briand and B. Picard, 2004. Muscle-specific metabolic, histochemical and biochemical responses to a nutritionally induced discontinuous growth path. Animal Science, Volume 204, 79-59.

Chrystall, B., 1994. Meat texture measurement. In: Pearson, A.M. and T.R. Dutson (eds.), Quality attributes and their measurement in meat, poultry, and fish products. Chapman & Hall, Glasgow, U.K.

Downey, G. and K.I. Hildrum, 2004. Analysis of meats. In: Roberts, C.A. (ed.), Near Infrared spectroscopy in Agriculture, Agronomy Monographs, Volume 44, ASA CSSA SSSA, Madison, WI, USA, 599-632.

Dransfield, E., J.F. Martin, D. Bauchart, S. Abouelkaram, J. Lepetit, J. Culioli, C. Jurie and B. Picard, 2003. Meat quality and composition of three muscles from French cull cows and young bulls. Animal Science, Volume 76, 387-399.

Dufour, E. and J.P. Frencia, 2001. Les spectres de fluorescence frontale - Une empreinte digitale de la viande. Viandes et Produits Carnés, Volume 22, 9-14.

Dufour, E. and A. Riaublanc, 1997. Potentiality of spectroscopic methods for the characterisation of dairy products, I - Front face fluorescence study of raw, heated and homogenised milks. Le Lait, Volume 77, 657-670.

Egelandsdal, B., G. Dingstad, G. Togersen, F. Lundby and O. Langsrud, 2005. Autofluorescence quantifies collagen in sausage batters with a large variation in myoglobin content. Meat Science, Volume 69, 35-46.

Elvidge, G., 2006. Microarray expression technology: from start to finish. Pharmacogenomics, Volume 7, 123-134.

Forrest, J.C., C.H. Kuei, M.W. Orcutt, A.P. Schinckel, J.R. Stouffer and M.D. Judge, 1989. A review of potential new methods of on-line pork carcass evaluation. Journal of Animal Science, Volume 67, 2164-2170.

Forrest, J.C., M.T. Morgan, C. Borggaard, A.J. Rasmussen, B.L. Jespersen and J.R. Andersen, 2000. Development of technology for the early post mortem prediction of water holding capacity and drip loss in fresh pork. Meat Science, Volume 55, 115-122.

Fortin, A. and D.P. Raymond, 1988. The Use of the Electrical Characteristics of Muscle for the Obgective. Detection of PSE and DFD in Pork Carcasses Under Commercial Conditions. Canadian Institute Food Science Technology Journal, Volume 21, 260-265.

Friedman, M., 1937. The use of ranks to avoid the assumptions of normality implicit in the analysis of variance. Journal of American Statistical Association, Volume 32, 675-701.

Garrido, M.D., S. Banon, J. Pedauye and J. Laencina, 1994. Objective Meat Quality Measurements of Ham - A Practical Classification Method on the Slaughterline. Meat Science, Volume 37, 421-428.

George, M.H., 1998. Palatability Assurance Critical Control Points for beef. Ph.D. Dissertation.

George, M.H., J.D. Datum, H.G. Dolezal, J.B. Morgan, J.W. Wise, C.R. Calkins, T. Gordon, J.O. Reagan and G.C. Smith, 1997. Comparison of USDA Quality Grade with Tender. Tec for the assessment of beef palatability. Journal of Animal Science, Volume 75, 1538.

Grunert, K.G., L. Bredahl and K. Brunso, 2004. Consumer perception of meat quality and implications for product development in the meat sector - a review. Meat Science, Volume 66, 259-272.

Harris, J., J.W. Savell, G.C. Smith and H.R. Cross, 1992. Armour Tenderometer readings and beef tenderness. Mimeograph Report. Meats and Muscle Biology Section, Department of Animal Science, Texas A&M University, College Station, TX, USA, 1-17.

Hildrum, K. I., J.P. Wold, V.H. Segtnan, J.P. Renou and E. Dufour, 2006. New spectroscopic techniques for on-line monitoring of meat quality. In: Di Leo, M., L. Nollet and F. Toldra (eds.), Advanced Technologies for Meat Processing, CRC Press, Boca Raton, FL, USA, 87-129.

Hocquette, J.F., 2004. Analogies for understanding statistics. Advanced in Physiology Education, Volume 28, 124-125.

Hocquette, J.F. 2005. Where are we in genomics? Journal of Physiology and Pharmacology, Volume 56, Suppl. 3, 37-70.

Hocquette, J.F., I. Cassar-Malek, A. Listrat and B. Picard, 2005a. Current genomics in cattle and application to beef quality. In: Hocquette, J.F. and S. Gigli (eds.), Indicators of milk and beef quality. EAAP Publication 112, Wageningen Academic Publishers, The Netherlands, 65-79.

Hocquette, J.F., I. Ortigues-Marty, D.W. Pethick, P. Herpin and X. Fernandez, 1998. Nutritional and hormonal regulation of energy metabolism in skeletal muscles of meat-producing animals. Livestock Production Science, Volume 56, 115-143.

Hocquette, J.F., R.I. Richardson, S. Prache, F. Médale, G. Duffy and N.D. Scollan, 2005b. The future trends for research on quality and safety of animal products. Italian Journal of Animal Science, Volume 4, Suppl. 3, 49-72.

Hocquette, J.F., and S. Gigli 2005. Indicators of milk and beef quality. EAAP Publication 112, Wageningen Academic Publishers, Wageningen, The Netherlands, 464 p.

Hocquette, J.F., B. Picard, G. Trillat, J. Normand, A. Boissy and J.Culioli J., 2001. Relations entre caractéristiques des fibres musculaires et indicateurs de qualité de la viande dans le cas du muscle *longissimus thoracis* de taurillon limousin. Rencontres autour des Recherches sur les Ruminants, Volume 8, 53-56.

Isaksson, T., L.P. Swendsen, R. Taylor, S.O. Fjæra and P.O. Skjervold, 2001. Non-destructive texture measurements of farmed Atlantic salmon, using Near Infrared Reflectance spectroscopy. Journal of the Science of Food and Agriculture, Volume 82, 53-60.

Isaksson, T.; B.N. Nilsen, G. Togersen, R.P. Hammond and K.I. Hildrum, 1996. On-line, proximate analysis of ground beef directly at a meat grinder outlet. Meat Science, Volume 43, 245-253.

Jaud, D., K. Weisse, K.H. Gehlen and A. Fischer, 1992. pH and conductivity - Comparative measurements on pig carcases and their relationship to drip loss. Fleischwirtschaft, Volume 72, 1416-1418.

Jenkins, T.G., K.A. Leymaster and M.D. MacNeil, 1995. Developement and evaluation of a regression equation of prediction for fat-free soft tissue in heterogenous populations of cattle. Journal of Animal Science, Volume 73, 3627-3632.

Jeremiah, L.E., L.L. Gibson, J.L. Aalhus and M.E.R. Dugan, 2003. Assessment of palatability attributes of the major beef muscles. Meat Science, Volume 65, 949-958.

Johnston, D.J., A. Reverter, D.M. Ferguson, J.M. Thompson and H.M. Burrow, 2003. Genetic and phenotypic characterisation of animal, carcass, and meat quality traits from temperate and tropically adapted beef breeds. 3. Meat quality traits. Australian Journal of Agricultural Research, Volume 54, 135-147.

Karoui, R.; B. Martin and E. Dufour, 2005. Potentiality of front-face fluorescence spectroscopy to determine the geographic origin of milks from Haute-Loire department (France). Le Lait, Volume 85, 223-236.

Kikugawa, K. and M. Beppu, 1987. Involvement of Lipid Oxidation-Products in the Formation of Fluorescent and Cross-Linked Proteins. Chemistry and Physics of Lipids, Volume 44, 277-296.

Lakowicz, J.R., 1983. Principles of Fluorescence Spectroscopy. Plenum Press, New York, USA.

Lebecque, A., A. Laguet, M. Chanonat, S. Lardon and E. Dufour, 2003. Joint analysis of sensory and instrumental data applied to the investigation of the texture of Charolais meat. Sciences des Aliments, Volume 23, 172-175.

Lehnert, S.A., Y.H. Yang, S.H. Tan and A. Reverter, 2006. Gene expression-based approaches to beef quality research. Australian Journal of Experimental Agriculture, Volume 46, 165-172.

Lepetit, J. and J. Culioli, 1994. Mechanical properties of meat. Meat Science, Volume 36, 203-237.

Lepetit, J. and C. Hamel, 1998. Correlations between successive measurements of myofibrillar resistannce of raw *longissimus dorsi* muscle during ageing. Meat Science, Volume 49, 249-254.

Lepetit, J., A. Grajales and R. Favier, 2000. Modelling the effect of sarcomere length on collagen thermal shortening in cooked meat: consequence on meat toughness. Meat Science, Volume 54, 239-250.

Lepetit, J., P. Salé, R. Favier and R. Dalle, 2002. Electrical impedance and tenderisation in bovine meat. Meat Science, Volume 60, 51-62.

Listrat, A. and J.F. Hocquette, 2004. Analytical limits of total and insoluble collagen content measurements and of type I and III collagens analysis by electrophoresis in bovine muscles. Meat Science, Volume 68, 127-136.

Listrat, A., C. Jurie, I. Cassar-Malek, L. Bouhraoua, B. Picard, D. Micol and J.F. Hocquette, 2001. Effect of grass feeding on muscle characteristics of finishing Charolais steers. Polish-French Symposium Animal and growth development: regulatory mechanisms. Reproduction Nutrition Development, Volume 42, 508.

Listrat, A., A.L. Pissavy, C. Jurie, C. Lethias, D.W. Pethick and, J.F. Hocquette, 2006. The relationship between collagen characteristics in muscle and marbling of beef. In: Troy, D., R. Pearce, B. Byrne and J. Kerry (eds.), Proceeding of the 52nd International Congress of Meat Science and Technology, Harnessing and exploiting global opportunities. Wageningen Academic Publishers, the Netherlands, 261-262.

Lu, R., Y.R. Chen, M.B. Solomon and B.W. Berry, 1998. Tensile properties and Warner-Bratzler tenderness measurement of raw and cooked beef. Transactions of the American Society of Agricultural Engineers, Volume 41, 1431-1439.

Luning, P.A., W.J. Marcelis and W.M.F. Jongen, 2002. Food quality management. A technico-managerial approach. Wageningen Pers., Wageningen, The Netherlands.

Maltin, C., D. Balcerzak, R. Tilley and M. Delday, 2003. Determinants of meat quality: tenderness. Proceeding of the Nutrition Society, Volume 62, 337-347.

Marshall, D.M., 1999. The genetics of meat quality. In Fries, R. and A. Ruvinsky (eds.), The genetics of cattle, CABI, Wallingford, UK, p. 605.

Masuyama, G. and S. Miura, 1962. Handbook for Sensory Tests in Industry. Juse Publishers, Tokyo, Japan.

McCormick, R.J., 1999. Extracellular modifications to muscle collagen: implications for meat quality. Poultry Science, Volume 78, 785-791.

Meilgaard, M., G.V. Civille and B.T. Carr, 1991. Sensory Evaluation Techniques. CRC Press, Boca Raton, FL, USA.

Melton, S.L., 1983. Methodology for Following Lipid Oxidation in Muscle Foods. Food Technology, Volume 37, 105.

Monin, G., 1998. Recent methods for predicting quality of whole meat. Meat Science, Volume 49, S231-S243.

Moser, D.W., R.M. Thallman, E.J. Pollak, M.E. Dikeman, C.A. Gill, S.R. Koontz, T.R. Holm and E.W. Dressler, 2004. Meeting consumer demands through genetic selection: the NCBA Carcass Merit Project. Proceeding BIF Annual Convention, Sioux Falls, SD, USA, 42-44.

Mullen, A.M., P.C. Stapleton, D. Corcoran, R.M. Hamill and A. White, 2006. Understanding meat quality through the application of genomic and proteomic approaches. Meat Science, Volume 74, 3-16.

Mullen, A.M. and D.J. Troy, 2005. Current and emerging technologies for the prediction of meat quality. In: Hocquette, J.F. and S. Gigli (eds.), Indicators of milk and beef quality. EAAP Publication 112, Wageningen Academic Publishers, The Netherlands, 179-190.

Munck, L., 1989. Fluorescence in Food Analysis. Longman Group UK Limited, Essex, UK.

Picard, B, L. Lefaucheur, C. Berri and M. Duclos., 2002. Muscle fibre ontogenesis in farm animal species. Reproduction Nutrition Development, Volume 42, 415-431.

Pliquett, F., U. Pliquett, L. Schönberlein and K. Freywald, 1995. Impedanzmessungen zur Charakterisierun der Fleischbeschaffenheit. Fleischwirtschaft, Volume 75, 496-498.

Purslow, PP., 2005. Intramuscular connective tissue and its role in meat quality. Meat Science, Volume 70, 435-447.

Ramanujam, N., 2000. Fluorescence Spectroscopy In Vivo. In: Meyers, R. A. (ed.), Encyclopaedia of Analytical Chemistry, John Wiley & Sons Ltd, Chichester, UK, 20-56.

Renand, G., 1988. Variabilité génétique de la croissance musculaire et conséquences sur les qualités de la viande chez les bovins. INRA Production Animal, Volume 1, 115-121.

Renand, G., C. Larzul, E. Le Bihan-Duval and P. Le Roy, 2002. L'amélioration génétique de la qualité de la viande dans les différentes espèces: situation actuelle et perspectives à court et moyen terme. INRA Production Animal, Volume 16, 159-173.

Renand, G., B. Picard, C. Touraille, P. Berge and J. Lepetit, 2001. Joint variability of meat quality traits and muscle characteristics of young Charolais beef cattle. Meat Science, Volume 59, 49-60.

Rhee, M.S., T.L. Wheeler, S.D. Shackelford and M. Koohmaraie, 2004. Variation in palatability and biochemical traits within and among eleven beef muscles. Journal of Animal Science, Volume 82, 534-550.

Salé, P., 1972. Appareil de Détection des Viandes Décongelées par Mesure de Conductance Electrique. Bulletin de l'Institut International du Froid, Volume 2, 265-275.

Salé, P., 1976. The electrical impedance of meat. 2ème Congrès International d'Impédance Bioélectrique, Lyon, France, 347-355.

Schmitten, F., K.H. Schepers, H. Jüngst, U. Reul and A. Festerling, 1984. Fleischqualität beim Schweine. Untersuchungen zu deren Erfassung. Fleischwirts, Volume 64, 1238-1242.
Schwarze, H., 1997. Continuous fat analysis in the meat industry. Process Control and Quality, Volume 9, 133-138.
Shoemaker, J., 2006. Statistical challenges with gene expression studies. Pharmacogenomics, Volume 7, 511-519.
Skjervold, P.O.; R.G. Taylor, J.P. Wold, P. Berge, S. Abouelkaram, J. Culioli and E. Dufour, 2003. Development of intrinsic fluorescent multispectral imagery specific for fat, connective tissue, and myofibers in meat. Journal of Food Science, Volume 68, 1161-1168.
Stolowski, G.D., B.E. Baird, R.K. Miller, J.W. Savell, A.R. Sams, J.F. Taylor, J.O. Sanders and S.B. Smith, 2006. Factors influencing the variation in tenderness of seven major beef muscles from three Angus and Brahman breed crosses. Meat Science, Volume 73, 475-483.
Swantek, P.M., J.D. Crenshaw, M.J. Marchello and H.C. Lukaski, 1992. Bioelectrical impedance: a nondestructive method to determine fat-free mass of live market swine and pork carcasses. Journal of Animal Science, Volume 70, 169-177.
Swatland, H.J., 1987. Remote monitoring of postmortem metabolism in pork carcasses. Martinus Nijhoff Publisher, Dordrecht, The Netherlands, p. 143-163.
Swatland, H.J., 1987. Measurement of the Gristle Content in Beef by Macroscopic Ultraviolet Fluorometry. Journal of Animal Science, Volume 65, 158-164.
Swatland, H.J., 1991. Evaluation of Probe Designs to Measure Connective-Tissue Fluorescence in Carcasses. Journal of Animal Science, Volume 69, 1983-1988.
Swatland, H.J., 1993. An Anomaly in the Effect of Temperature on Collagen Fluorescence in Beef. Food Research International, Volume 26, 271-276.
Swatland, H.J., 1996. Connective tissue distribution patterns in beef detected by ultraviolet fibre optics. Food Science and Technology – Lebensmittel-Wissenschaft & Technologie, Volume 29, 272-277.
Thallman, R.M., D.W. Moser, E.W. Dressler, L.R. Totir, R.L. Fernand, S.D. Kachman, J.M. Rumph, M.E. Dikeman and E.J. Pollak, 2003. Proceedings 8th BIF Genetic Prediction Workshop Molecular approaches to genetic improvement, Kansas City, MO, USA 70-90.
Thompson, J.M., 2002. Managing meat Tenderness. Meat Science, Volume 60, 365-369.
Tilgner, D.J., 1957. Analiza organoleptyczna zywnosci, Warszawa: Wydawnictwo przemyslu Lekkiego I Spozywczego.
Timm, R.R., J.A. Unruh, M.E. Dikeman, M.C. Hunt, T.E. Lawrence, Jr. J.E. Boyer and J.L. Mardsen, 2003. Mechanical measures of uncooked beef longissimus muscle can predict sensory panel tenderness and Warner-Bratzler shear force of cooked steaks. Journal of Animal Science, Volume 81, 1721-1727.
Togersen, G.; T. Isaksson, B.N. Nilsen, E.A. Bakker and K.I. Hildrum, 1999. On-line NIR analysis of fat, water and protein in industrial scale ground meat batches. Meat Science, Volume 51, 97-102.
Voisey, P.W., 1976. Engineering Assessment and Critique of Instruments Used for Meat Tenderness Evaluation. Journal of Texture Studies, Volume 7, 11-48.
Von Seggern, D.D., C.R. Calkins, D.D. Johnson, J.E. Brickler and B.L. Gwartney, 2005. Muscle profiling: Characterizing the muscles of the beef chuck and round. Meat Science, Volume 71, 39-51.
Warner, K.F., 1928. Progress report on the mechanical tenderness of meat. Proceeding of the American Animal Production, Volume 21, 114-125.
Weatherell, C., A. Tregear and J. Allinson, 2003. In search of the concerned consumer: UK public perceptions of food, farming and buying local. Journal of Rural Studies, Volume 19, 233-244.
Wheeler, T.L., S.D. Shackelford and M. Koohmaraie, 1998. MARC beef classification system: Objective evaluation of beef tenderness of beef tenderness and cutability. ARS, USDA, Washington, DC, USA.
Wold, J.P. and M. Mielnik, 2000. Nondestructive assessment of lipid oxidation in minced poultry meat by autofluorescence spectroscopy. Journal of Food Science, Volume 65, 87-95.
Wold, J.P.; M. Mielnik, M.K. Pettersen, K. Aaby and P. Baardseth, 2002. Rapid assessment of rancidity in complex meat products by front face fluorescence spectroscopy. Journal of Food Science, Volume 67, 2397-2404.
Yin, D.Z., 1996. Biochemical basis of lipofuscin, ceroid, and age pigment-like fluorophores. Free Radical Biology Medicine, Volume 21, 871-888.

Methodologies to evaluate meat quality in small ruminants

Carlos Sañudo[1], M.M. Campo[1], J.L. Olleta[1] M. Joy[2] and R. Delfa[2]
[1]Unit of Animal Production, University of Zaragoza, 50.013 Zaragoza, Spain
[2]Unit of Technology in Animal Production, C.I.T.A., Apdo. 727, 50.080 Zaragoza, Spain

Abstract

The assessments of meat quality and different factors that can influence it in small ruminants have been discussed. The muscle *Longissimus dorsi* is the most used muscle to study meat quality and a sampling proposal within the muscle is shown. Meat quality parameters include pH, water holding capacity, instrumental texture, colour, sensory analyses through a panel test and consumers and chemical characteristics. Special emphasis has been put on genotype, sex, age, weight and feeding as factors that affect these characteristics

Keywords: techniques, meat quality, lamb, goat

Introduction

Small ruminants have been one of the first domesticated species and from their origins have produced multiple benefits to humans. In the last two centuries, world sheep production has become more and more specialised upon geographical areas. Thus, big areas from America and Australia are dedicated mainly to fine wool, the Middle East and large areas of Asia are orientated towards skim and coarser wool, North Europe has specialised in meat and the Mediterranean keeps a clear vocation in milk production. There are regions where double or triple aptitude breeds have been developed. However, goat has maintained its natural adaptation to the environment even when in some geographical regions dairy breeds are increasing their population whereas in others, double aptitude local breeds are only dedicated to meat production.

On the other hand, recent changes in consumer aptitudes, together with the increment in artificial fibre industries and human world population, some crisis related with the international wool market and some increments in lamb prices, have provoked that lamb meat has increased its relative importance inside the global sheep income. Also, sheep production systems vary considerably across the world, and reflect the different local environmental conditions, which determine, to a large extent, breeds, housing, levels of intensification and, at the end, local market requirements and qualities. Lamb and goat meat world production is around 12 million tonnes (3.7 millions in goat; FAOSTAT, 2001). Small ruminant meat consumption is around 2 kg per capita with large variations between continents or even between countries or regions within the same country. Thus, sheep consumption is around 7 kg in Oceania or 0.7 kg in North America. With 1.9 million Tm in 2002, the international trade of sheep represents a small percentage of all the international meat trade. However, among all the meats, sheep meat remains the most internationally traded (16 % of total world production is exported), which explains the potential existence of different lamb products in the same market. Goat meat is mainly locally consumed, and mainly in rural areas, thus only 0.5% of the total meat produced is marketed internationally, but a more presence in the market is required to assure the future of species in a competitive way.

Those changes in small ruminant aptitudes and production requirements have created a higher interest for meat production including its quality. This interest from producers and the industry should be reflected in a proportional higher interest from the researches towards sheep and goat, where it exists, in general, important lacks of knowledge (Devendra, 2000; Lebbie, 2000) and, in particular, in lamb and kid meat Science.

On the other hand, results of studies from pigs or beef are many times applied to small ruminants, which could be a mistake because of the high amount of specificity in these species, including meat (Sierra and Sañudo, 1987; Morand-Fehr and Lebbie, 2004). It seems clear that, to improve international research, making as comparable as possible the results from different authors, a common language and methodologies should be employed.

Some attempts to use co-ordinated methods to evaluate meat quality have been developed in specific analytical techniques or in beef (Boccard, *et al.*, 1981, Honikel, 1997 and 1998), or ruminants in general (Cañeque and Sañudo, 2005) with more or less followers, but up to date, considering the last 2 or 3 decades, there is not a specific proposal to evaluate lamb or kid meat quality in all Europe. It could be considered that it is always difficult to put in agreement different laboratories and researchers, most of the times for differences in availability of equipment or traditional work routines and not for disagreements between researchers or techniques, because not discussion at this time has been happened. To analyse the state of the art, to make some proposals to evaluate small ruminant quality and some factors that affect this quality and to show some numeric results, with the aim of improving the efficacy and efficiency of meat research in Europe, are the objectives of the present paper.

Sampling

At the moment, we do not know any work that, at European level, analyses a specific muscle as representative of the whole animal in order to study meat quality. The chosen muscle should be homogeneous as a whole, and large enough to provide the necessary samples for all the analysis. This muscle should be easily separated from the carcass and should not depreciate the carcass too much in order to make the studies less expensive. Finally, it would be desirable that it could be defined by all its characteristics, as the representative muscle of the meat quality that carcass can offer.

Even though this magic muscle does not exist or we have not been able to discover it; probably, in small ruminants, *longissimus dorsi* could be the best muscle, at least considering its size and relative homogeneity.

Sañudo *et al.* (2005) explained that probably it will be difficult that just one muscle may represent all the requirements, especially because there are different muscular physiological models according to the typology of their fibres (Talmat *et al.*, 1986). For this reason, it seems logical that instead of limiting the study to one muscle, several of them should be studied at the same time, but this 'ideal' has many practical problems, because it would be very expensive, due to the human and material required.

A revision of 58 works on sheep meat published in Meat Science from January 2001 to May 2005 (Sañudo *et al.*, 2005), comprised the studies on fourteen different muscles: in 36.2% of the papers only one muscle was studied, 37.9% studied two muscles, 22.5% studied more than two muscles and two papers did not indicate the sample used. In these papers the main muscles studied were: *Longissimus dorsi* (*thoracis* and/or *lumborum*) (93.1% of the cases), *Semimembranosus* (21.4%), *Semitendinosus* and *Biceps femoris* (12.1%) and *Triceps brachii* (6.9%). In addition, *Longissimus* was named in 34.5% of the occasions *Longissimus dorsi* and in the rest *thoracis* and/or *lumborum*. This muscle presents several qualities that make it especially interesting: it is a long muscle, easily dissected from the carcass, presenting a convenient slicing, with high commercial value (first category), and therefore, representative of the carcass commercial value as well as an average metabolism. However, the main disadvantages are its high price and the direction of its fibres, oblique to the muscle central axis what makes difficult to carry out specific analytical techniques. On the other hand, its qualitative characteristics may change along its extension, especially when comparing separated parts, while not so much when slices are successive or near. This makes necessary to limit the area or areas within the muscle to be used for every analysis. It is included a sampling proposal that tries to answer and standardise sampling to different common meat quality analysis (Figure 1). Lamb *longissimus* muscle samples should be taken between the third thoracic vertebra (T3) and the fifth lumbar vertebra (L5). Thus, the proposal would include the lumbar part for sensory analysis, the last three dorsal vertebrae for texture analysis and the rest (T3-T10) of dorsal vertebrae for chemical analysis and water holding capacity. Measurements of pH and colour could be made on the limit between the dorsal and lumbar regions.

The chosen half carcass will be the left one, although in some occasions it may be necessary to use both sides. Preferably, the *longissimus dorsi* muscle should be dissected from the carcass 24 h after slaughtering, maintaining all the hygienic procedures, including the use of mask, globes, whitecoat and hats.

Samples will be preferably sliced keeping subcutaneous fat and surrounding muscles from the area, for a better protection during the storing period. They will be *vacuum* packaged carefully, paying special attention to avoid modifications of their natural shape.

The time of recommended ageing for the analysis will depend on the objectives of the study. As a general rule, 3 or 6 days could be reasonable, taking into account that long ageing periods will decrease differences between treatments due to the tendency of the final product to homogenise quality (Martínez Cerezo, 2005a-c). The freezing time should not be very long (no more than six months), though one-year-old freezing could be acceptable. In any case, a minimum time of freezing should be required for all the samples (one week), when there are at least one-month-old samples of the same experiment.

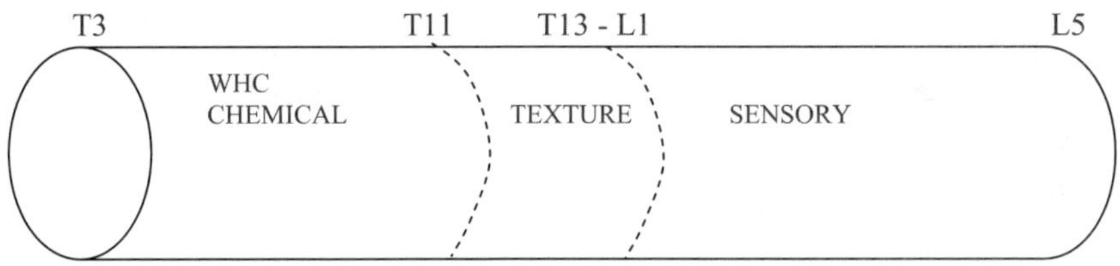

T = thoracic vertebra; L = lumbar vertebra.

*Figure 1. Basic proposal for sampling in small ruminants (Longissimus dorsi, thoracis and lumborum, one side disposability). *pH and physical colour could be measured at T13-L1 level.*

pH

The pH influences the meat quality. Post-mortem pH is determined by the quantity of lactic acid produced from glycogen during anaerobic glycolisis. This process will be interrupted if the muscle glycogen reserves are run down due to fatigue, starvation or fear before slaughtering (Lawrie, 1998). Its value is represented by a 14 points scale where the intermediate section (pH = 7) corresponds to neutral pHs, zero means the higher acidity degree and 14 the higher basicity degree.

The muscles of live animals present a neutral pH, with normal values between 7.0 and 7.3. Forty minutes after the slaughter, pH decreases slightly below 7.0, reaching minimum final values between 5.4 and 5.7 at 24 hours after the slaughter at standard temperatures (Wheeler and Koohmaraie, 1994); remaining practically unalterable along ageing as Vergara and Gallego (2000) stated for sheep for a period between one and 14 days of ageing. However, slight pH changes were detected in animals electrically stunned before slaughtering. The rate of the post-mortem declined and the pH level the muscle reached ultimately influenced the storage life, attractiveness to purchasers, and eating qualities of meat (Bray *et al.,* 1994). Pale, soft and exudative pork is produced when pH levels fall rapidly, and dark, firm and dry beef and pork are produced when ultimate pH levels are high (Pearson and Young, 1989). While lamb meat may not exhibit effects as dramatic as observed in pork and beef, the pH evolution after slaughtering is going to have a deep effect on the organoleptic (colour, juiciness, flavour) and technological properties of meat (water holding capacity and shelf life).

The pH potentiometric measurement is based on the recording of the difference of electrical potential between a measuring electrode and a reference electrode at a specific temperature. There is a dependence from the pH with respect to the temperature in –0.006 units of pH/ºC for a temperature level from zero to 20 ºC in the case of sheep (Jansen, 2001). The pH measurement must be done at 24 h post-mortem, recommended at least on the *longissimus dorsi* from the left half carcass, and between the fourth and fifth lumbar vertebrae. The electrode should be introduced in depth into the muscle, and pH recorded when the lecture is stable.

pH is affected by:

- *Genotype*. The effect of the breed on the pH does not seem very important (Martínez-Cerezo *et al.*, 2005a). Teixeira *et al.* (2005) concluded that difference between breeds could be attributed to the stress rather than breed.
- *Sex* had no effect on the pH measured 24 h post-mortem (Velasco *et al.*, 2000; Teixeira *et al.*, 2005). However, when pH was measured 1 h after slaughter, there is a small tendency for males to have a higher pH (Vergara and Gallego, 1999; Fogarty *et al.*, 2002).
- *Weight and age*. Age has little effect on pH (Jaime *et al.*, 1992) and the slaughter weight have a certain positive tendency to increase the pH value (Beriain *et al.*, 2000; Diaz *et al.*, 2002) or no effect on pH (Vergara *et al.*, 1999; Velasco *et al.*, 2000; Martínez-Cerezo *et al.*, 2005a).
- *Fatness degree.* Lambs with different fatness degree has the same stress susceptibility and muscle glycogen content, and the pH values are practically the same in all classes of fatness (Sañudo *et al.*, 2000b).
- *Feeding and production system.* The production system has no effect on pH values and their variations (Diaz *et al.*, 2002). Velasco *et al.* (2004) concluded that the type of feed did not affect the pH at any of the times studied, nor did it influence the drop in pH values throughout carcass refrigeration. However, Priolo *et al.* (2002) indicate that the last pH in grass lambs tended to be higher than in stall lambs.

Water Holding Capacity (WHC)

The WHC was described by Hamm (1960) as the ability of meat to hold its constitutive water during the application of strange forces or a treatment. This property affects qualitative aspects (retention of vitamins, minerals, salts) and quantitative ones (volume of retained water) of both, fresh meat and meat products (Horcada, 1996). This last author indicates that Gault (1985) and Honikel (1991) consider that the WHC of the meat depends more than 80% on the pH final value and on the muscle pH fall. Therefore, it seems that there is a lineal relationship between both parameters (final value and pH fall) that has been largely studied by Bouton *et al.* (1971) on sheep.

The WHC is usually determined by the estimation of drip loss (natural exudation of meat), thawing loss (exudation after freezing and defrost) or cooking loss (exudation after heating); and by evaluating the quantity of expressible juice (exudation after application of external forces such as compression, centrifugation or suction). In laboratories, the most commonly used is the method of Grau and Hamm (1953). A variation of this method is the modified compression technique described by Sierra (1973). Here there are described two of the most common methods used (Honikel, 1998):

Drip loss is based on the fluid from a muscle that will drain from the surface under gravity if the viscosity of the fluid is low enough and capillary forces do not retain it. The methods chosen for measuring drip loss must conserve the integrity of the muscle before sampling in order to avoid external forces other than gravity. Orientation of the fibres with respect to cut should be taken into consideration. Surface evaporation has to be prevented and the method supporting the meat pieces should minimize tension (suspended from above) or compression (supported from below). The equipment required is a balance, a sealable, water impermeable bag, sample support that allows the escape of fluid and a temperature-controlled environment.

Cooking loss. During heating, the meat protein denature and causes structural changes such as the destruction of cell membranes, transverse and longitudinal shrinkage of muscle fibres, the aggregation of sarcoplamic proteins and shrinkage of the connective tissue. Cooking conditions must be defined and controlled. The equipment required is: a balance, a temperature-controlled water bath, thin walled polyethylene bags and thermocouples to allow temperature recording in the centre of each sample.

WHC is affected by some factors like:

- *Genotype.* Breed effects on meat quality do not seem to be very important (Sañudo *et al.*, 1998; Dransfield *et al.*, 1990; Rousset-Akrim *et al.*, 1997; Santos Silva *et al.*, 2002). The results of breed are uncertain, depending on the compared breed; there are works where it does affect while it does not in others (Martínez-Cerezo, 2005a).
- *Sex.* Sañudo *et al.* (1998a) indicate that the differences between sexes in meat quality are not, in general, very important. Several authors (Dransfield *et al.*, 1990) had not found differences between males and wethers in the pH, WHC or colour. However, Velasco *et al.* (2000) explain that the WHC was influenced by gender being this value greater in males.
- *Weight and age.* Studies of the effect of age and weight on WHC showed controversial results. The age has no effect on WHC. The effect of slaughter weight is less clear. While Díaz *et al.* (2002), and Velasco *et al.* (2004) reported that WHC was not significantly influenced by carcass weight, Russo *et al.* (2003) observed that heavier carcasses have a better WHC for raw meat.
- *Fatness degree.* Fatness degree has no effect on WHC or cooking losses (Sañudo *et al.*, 2000b).
- *Feeding and production system.* Santos Silva *et al.* (2002) indicate that the feeding system was not associated with important changes in meat quality. Díaz *et al.* (2002) and Velasco *et al.* (2004) concluded that the production system had no effect on WHC. Nevertheless, meat from lambs reared on pasture presented lower WHC than meat for lambs concentrate-fed.

Texture

For decades, consumers considered tenderness as the most important quality attribute of meat. In a review of factors influencing consumption, selection and acceptability of meat, Jeremiah (1982) concluded that the most common cause of unacceptability in beef was toughness and that it was a usual problem in pork and lamb. Texture is a complex concept that includes many attributes, being the most important toughness or tenderness. Therefore, it is not possible to make a reference to the global texture of a specific product, it is necessary to talk about the different textural characteristics or attributes of this food. In meat, texture is related to its different structural components and could be evaluated both sensorial and instrumentally.

Several authors have studied the influence of pre-slaughter factors on lamb texture. However, texture is also influenced to a large extent by slaughter and post-slaughter conditions. Breed, individual factors, specific genes, weight, age, pH, muscle type, carcass chilling, electrical stimulation, preservation method, cooking, the contractile state, the degree of post mortem degradation of the myofibrillar component, and connective tissue properties mainly influence lamb texture. (Bouton and Harris, 1972; Field *et al.*, 1978; Jeremiah, 1982; Devine *et al.*, 1993; Devine and Graafhuis, 1995; Koohmaraie *et al.*, 1995; Sañudo *et al.*, 1998a,b; Berge *et al.*, 2000; Purslow, 2005).

Breed, as a source of variation, is a complex factor, as the results will vary depending on the comparison criterion used: equal live or carcass weight, equal age, equal degree of maturity or equal percentage of adult live weight. The most significant differences in texture can be justified by differences in precociousness or in the degree of muscularity. Thus, the most precocious breeds and the meat purpose ones are more tender, although for different reasons, than breeds with intermediate characteristics.

Tenderness is affected by some factors as:

- *Callipyge gen* causes hypertrophy in sheep and is associated with an enhanced growth, less fat and excessively tough meat.
- *Age*. It has been suggested that the effect of age on the shear force and tenderness of lamb is relatively small, but young animals would be more tender, as they possess a more soluble collagen. However, other researchers indicate that shear force values or compression values are higher for medium weights (10-11 kg) than those for heavier (13-14 kg) or smaller (7-8 kg) carcasses. Also, the relationships between intermediate pH and increased levels of toughness have been demonstrated in sheep. It has been attributed predominantly to direct effects of pH on the activity of the proteolytic enzymes, which degrade the myofibrillar structure of the muscle.
- *Post mortem effects*. Throughout, ageing meat is more tender. Electrical stimulation produces improvements in tenderness because accelerates the onset of *rigor mortis* and reduces cold-induced shortening. It is well documented that storage and subsequent ageing of meat at refrigerated temperatures post mortem results in a significant improvement in tenderness. A rapid carcass chilling produces the toughening of the meat. The effects of pre rigor freezing on the development of rigor mortis and tenderness have been known for many years. Thawing pre rigor frozen muscle results in thaw shortening but this can be prevented by holding, in the frozen state, the pre rigor frozen meat for long enough to allow full *rigor mortis* to develop. Different cooking conditions (type, times and temperatures) have a large effect on meat toughness. The changes in meat toughness can be explained by the effect of heating on connective tissue (changes in the mechanical properties of the perimysial connective tissue) and the heating denaturation of myofibrillar proteins, especially myosin. The amount, spatial distribution and composition of the connective tissue within muscle vary with muscle position in the carcass and with animal age. This has long been recognised to influence the tenderness of cooked meat. Crosslinking of collagen in older animals is generally considered to result in tougher meat, although definitive links between mature crosslink content and cooked meat toughness have been difficult to prove.

In many studies, some of these factors have been mixed, for example when high *versus* low energy diets want to be compared at the same slaughter weight and differences in age are not considered. This has lead to a considerable controversy in the literature that could be clarified by using many animals and large-scale studies in controlled conditions.

Instrumentally textural properties of meat have been studied with many different instruments and probes. Also, raw or different cooking or heating methods have been used prior to textural studies.

Shear force, using the Warner Bratzler (WBSF) probe (a triangular shape lame), is the most common method to evaluate meat toughness. The mechanical behaviour of meat is complex, with a viscoelastic contribution, and its mechanical properties therefore depend on the rate at which strain occurs. Also, results could change upon temperatures, heating and posterior cooling, homogeneity of the sample and fibres direction in relation to the cutting axis. We recommend a water bath in a sealed package at 75 °C, until a 70 °C internal temperature is reached, cooling for 30 min in tap water and storage at room temperature until tested. Samples ranging between 80 and 150 grams depending of carcass size, should be divided in blocks of similar dimensions (square 100 mm^2 cross-section, 10x10 mm, 20 mm length with the longitudinal axis parallel to the fibre direction). As many good blocks as possible should be obtained. Also a crosshead speed of 150 mm/min and measurements logged at a rate of 20 data points/second are recommended.

Some of the myofibrillar resistance characteristics can be quantified by the sarcomere length and by tests at low mechanical strains as long as collagen fibres in the connective tissue remain in a more crimped state than myofibres, in raw meat (compression 20%), whilst the strength of connective tissue can been obtained at higher strains (compression 80%) (Lepetit 1991; Lepetit and Culioli, 1994). Sañudo *et al.* (2003) reported some differences on lambs from several breeds and production systems. There were differences in the *Longissimus dorsi* muscle in shear force, compression test (20% and 80%) and sarcomere length. The test was performed in longitudinal configuration, using a modified compression device that avoided transverse elongation of the sample, with applied strain perpendicular to muscle fibres and free strain parallel to muscle fibres. Samples were subjected to the shear test, at right angles to the muscle fibre axis, using a Warner-Bratzler device with a triangular shaped hole in the shear blade, mounted on an Instron machine (Table 1). Sarcomere length could be measured in 4x4x8 mm samples, fixed in 2.5% phosphate-buffered glutaraldehyde (Merck®), which produces a derivate shortening. The fibres, either single or in small groups, are carefully teased out and sarcomere length can be measured directly by light microscopy. Areas of fibres showing crimping must not be used for determination. A minimum of 20 reps from different fibres in the same muscle preparation should be used.

Colour

Meat colour is due to the concentration of pigments (myoglobin, hemoglobin), their chemical state and the light-scattering properties of the meat (Lawrie, 2002). But when meat is cooked, globin denatures to a degree that depends on the ultimate temperature and pH of the tissue, however the hemo portion remains intact. The visual appearance of the meat is an essential factor for consumer acceptance. In this way, meat is selected for its leanness, general appearance and colour. In Mediterranean countries, light colour is associated with meat from young animals, which is preferred, having a great influence on the price. On the other hand, there are other countries where darker meat is more easily accepted.

The objective measurements of the meat colour are generally taken using the CIE colour system measured using a Minolta spectrophotometer/colorimeter set on the L*, a*, b* system (Commission International de l'Eclairage, 1986, where L* measures relative lightness, a* relative redness and b* relative yellowness). Illuminant D65 is preferred over illuminant C with the observer of 10°.

Apart from species-dependent effects, individual factors and feeding system, colour variation is dependent on many intrinsic factors (sex, breed, endogenous antioxidants, age of animal, muscle type and metabolism, the ultimate pH of meat, rate of post-mortem pH decline, amount of intramuscular fat, oxygen consumption rate). Also, extrinsic factors (temperature, O_2 availability, light type exposure, packaging, type and growth of surface microorganisms, irradiation) are important (Renerre, 1986; Monin, 1989; Faustman, 1994; Sañudo *et al.*, 1998a; Jeremiah, 2000).

Table 1. Meat textural characteristics in different European commercial lambs: Warner-Bratzler test on cooked meat and pH (n=108 per lamb type) compression test and sarcomere length in raw meat (Sañudo et al.*, 2003).*

Lamb size and description	WBSF (kg)	Compression (N/cm^2)		Sarcomere length (μm)
		Stress 20%	Stress 80%	
Suckling lamb, Spanish	3.15 ^{efgh}	11.79 ^{c}	55.62 ^{ab}	1.53 ^{ab}
Suckling lamb, Greek	2.86 ^{bcdefg}	10.52 ^{bc}	69.72 ^{b}	1.63 ^{abc}
Light Spanish lamb, concentrates	2.43 ^{abcde}	6.50 ^{ab}	60.75 ^{ab}	1.60 ^{abc}
Light British lamb, grass	2.29 ^{abcd}	5.24 ^{a}	49.30 ^{ab}	1.68 ^{bc}
Light Greek lamb, concentrates	3.56 ^{ghij}	4.67^{a}	47.47 ^{ab}	1.56 ^{ab}
Light Italian lamb, concentrates	3.65 ^{hij}	7.74 ^{abc}	49.73 ^{ab}	1.47 ^{a}
Medium Greek lamb, grass	2.88 ^{cdefg}	4.96 ^{a}	57.34 ^{ab}	1.61 ^{abc}
Medium British lamb, grass	2.16 ^{abc}	4.96 ^{a}	52.89 ^{ab}	1.56 ^{ab}
Medium French lamb, concentrates	2.24 ^{abcd}	6.47 ^{ab}	53.13 ^{ab}	1.55 ^{ab}
Medium Greek lamb, grass + concentrates	2.44 ^{abcde}	6.66 ^{ab}	63.88 ^{ab}	1.53 ^{ab}
Medium Icelandic female lamb	1.72 ^{a}	5.00 ^{a}	49.02 ^{ab}	1.65 ^{abc}
Medium Icelandic male lamb	1.75 ^{a}	4.84 ^{a}	48.70 ^{ab}	1.72 ^{bc}
Heavy British lamb, grass	2.95 ^{defgh}	5.10 ^{a}	48.60 ^{ab}	1.55 ^{ab}
Heavy Italian lamb, grass	4.13 ^{j}	4.73 ^{a}	53.52 ^{ab}	1.77 ^{c}
Heavy British lamb, concentrates	2.76 ^{bcdef}	5.24 ^{a}	44.60 ^{a}	1.68 ^{bc}
Heavy Italian lamb, transhumance	3.98 ^{ij}	4.66 ^{a}	59.34 ^{ab}	1.47 ^{a}
Std.	1.21	3.04	13.29	0.13
F	63.02***	10.14***	5.58***	10.32***

$^{a\text{-}j}$ Different superscripts within columns, indicate significant differences (P<0.05).

Gender differences are not noticeable, in general, although females could be darker than males due to their greater precociousness and fatness when comparing both sexes at equal slaughter weight.

Slaughter live weight and breed influence meat colour, but it is more important the first effect than the second. As slaughter weight increases, meat lightness decreases. Meat has been shown redder at 20-25 kg than at 10-12 kg live weight. However, the increase from 20 to 30 kg did not affect redness, which suggests that the change in diet (milk to concentrate) probably has a greater effect on meat colour than carcass weight or age.

Colour is affected by:

- *Carcass weight* affects colour variables with a tendency to become darker, for the amount of pigments. But colour evolution may not be linear and at certain ages the colour may become stable or change at a greater speed than expected (Alexandrova *et al.*, 1996).
- *Milk feeding* until slaughter produces paler meat with less pigment, in very young animals, but, at 5 months of age, milk feeding does not result significantly in paler meat than roughage feeding, in part because lambs are quite tolerant to low iron diets (Okeudo *et al.*, 1994). These results of meat from suckling lambs (lighter and less red) could be due to the low iron content of ewe milk, leading to lower myoglobin synthesis in muscle (Lawrie, 1985). Also, meat from unweaned lambs shows less lightness (L* index) and redness (a lower a* index) than that from weaned animals, due to the lack of iron in milk. In relation to the yellowness index (b*) unweaned lambs exhibited higher values than weaned lambs (Table 2).

Table 2. Effect of breed, slaughter live weight (LW) and feeding system (FS) on lamb meat colour.

Breed	LW FS	10-12 kg Suckling	20-25 kg Concentrate	30-35 kg Concentrate	30-35 kg Grass	50 kg Grass
Churra[1]	L*	46-50	40-43	39-42		
	a*	8-11	14-16	14-16		
	b*	5-9	7-9	6-8		
Spanish Merino[2]	L*	43-46	37-42	38-41		
	a*	11-12	14-16	14-16		
	b*	5-7	6-10	6-8		
Rasa Aragonesa[3]	L*	42-46	37-44	37-40		
	a*	7-11	12-17	12-17		
	b*	5-9	6-12	6-8		
Lacha[4]	L*	46-49	40-44	40-43		
	a*	7-8	9-10	10-11		
	b*	8-11	6-8	6-8		
Manchega[5]	L*	45-51	45-50			
	a*	8-14	15-22			
	b*	6-10	7-9			
Bragançana/ Mirandesa[6]	L*	43-45	38-40			
	a*	15-16	16-18			
	b*	10-11	8-9			
Barbaresca[7]	L*	45-48				
	a*	9-11				
	b*	4-6				
Lacaune[8]	L*		44-48			
	a*		14-15			
	b*		6-9			
Ile de France[9]	L*			49-50	45-47	
	a*			8-14	8-10	
	b*			10-11	9-10	
Merino Branco[10]	L*			38-41		
	a*			15-18		
	b*			4-5		
British Crossbred lambs[11]	L*					37-39
	a*					15-17
	b*					6-7
Poll Dorset x Border Leicester x Merino[12]	L*					35-36
	a*					17-18
	b*					6-7

[1] Sañudo *et al.*, 1997; Martinez *et al.*, 2005a. [2] Alcalde *et al.*, 1999; Martinez *et al.*, 2005a. [3] Sañudo *et al.*, 1996; Sañudo *et al.*, 1998b; Alcalde *et al.*, 1999; Beriain *et al.*, 2000; Sañudo *et al.*, 2000; Martinez *et al.*, 2005b; [4] Beriain *et al.*, 2000; [5] Sañudo *et al.*, 1997; Alcalde *et al.*, 1999; Vergara and Gallego, 1999; Vergara *et al.*, 1999; Díaz *et al.*, 2003; Cañeque *et al.*, 2004; [6] Teixeira *et al.*, 2005; [7] Lanza *et al.*, 2006; [8] Alcalde *et al.*, 1999; [9] Priolo *et al.*, 2002; [10] Santos-Silva *et al.*, 2003; [11] Alcalde *et al.*, 1999; [12] Hopkins *et al.*, 2001.

- *Production system* is also important in terms of meat colour. Thus, meat of animals raised at pasture is generally darker than that of sheepfold animals fed concentrate, as lambs at pasture show a greater concentration of blood pigments in muscle than sheepfold lambs (Renerre, 1986).
- *Vitamin E* has a significant effect on the colour stability, extending the shelf life of lamb meat several days, depending directly on its concentration.
- *Post slaughter* conditions affect largely meat colour, thus greater ageing time develops an increase in the discoloration and browning with a rapid evolution of objective colour (L*, a*, b*). Freezing produces colour modifications, with a tendency to reduce L* and b*.
- *Muscle.* The muscle *rectus abdominis* displayed greater lightness (L*) and lower yellowness (b*) index than muscle *longissimus dorsi*. Nevertheless, *rectus abdominis* is a very thin muscle and some differences in colour may often go undetected compared to much thicker muscles such as *longissimus dorsi* (Díaz *et al.*, 2003; Lanza *et al.*, 2006).

Panel test

For countries with more lamb production and vocational export capacity, that have a well-developed meat chain, it is a very important and positive competitive advantage to know the characteristics, preferences and acceptability of their products in foreign markets. In Europe, there is a large variability in lamb production and subsequently in the culinary background of its citizens. In any case, this variability makes, a priori, a clear difference between North and Mediterranean countries, and even between consumers that has high or low lamb consumption levels. Thus, Mediterranean regions apparently have pale and delicately flavoured flesh, in contrast to the larger, fatter and more robustly flavoured types produced in northerly areas. Also, countries such as Spain or UK have relatively high lamb consumption whereas others, such as Germany, have very low rates. Historically, it has been shown that in areas with low meat consumption, sheep meat is generally disliked because of its strong flavour and odour. However, in areas where people consume a large amount of sheep meat, apparently flavour is more enjoyed (Griffin *et al.*, 1992; Font i Furnols *et al.*, 2006). That variability in tastes represents market opportunities to change or to diversify the offer. Nevertheless, in any case, it is not completely clear which type of lamb products respond to the demand.

Sensory analysis is an analytical methodology that uses human senses to describe and quantify meat attributes. Because of the high human perception variability, it is necessary a precise methodology in both assessors qualification and test methodology.

A standardised method of recruiting, selection and training should be used, and international standard procedures are available (ISO, ASTM, UNE, BSI, etc.). The recruiting process should consider which types of assessors we want to include in the panel, the decision mainly is between internal or external assessors. Both have several advantages and disadvantages and the final decision should be carefully taken. In our opinion, when possible, an external group has more advantages. At least, three times more people than the final number of assessors required should be initially recruited.

After recruiting, the most important and probably neglected part of the process must be initiated: the selection phase. At this time, psychological and physiological capacity should be considered in a preliminary selection period. To detect both, several tests are described, but usually the sensory capacity to recognise, describe and order is considered more important than psychological aptitudes such as character, capacity to concentration, motivation or interest. These latest aspects, together with the feeling between the future panellist and the panel leader, are especially important for the success of the sensory analysis. It is important to have a clear knowledge of the person (health, allergies, mental agility, creativity, intellectual capacity and so on), in order to initiate the process having more information to take the final decision on the election of the panel members. Some questionnaires are specifically described (Meilgaard *et al.*, 1991).

After this preliminary phase, in which only the subjects with visual or general sensory incapacity (colour blindness, anosmie, ageusia), or clearly inadequate (health, comprehension problems, any type of incompatibility, etc.) will have been eliminated, the specific selection should be start. Previously to this specific selection, it is important to initiate a short formative period, in which the physiology of senses, the basic rules of the sensory analysis and an elemental description of the products target of the future analyses should be shown. This sessions have the objective of homogenising a minimum of the previous knowledge of the candidates and, therefore, having more security of choosing those with higher sensorial aptitudes and not only those with previous formation (Guerrero, 2005).

Inside the specific selection program, tests related to sensory aptitudes (discriminative, memory, differentiate, descriptive, quantitative) and global capacities (compression, description, understanding, discriminating ability, etc.), should be used to get as much information as possible to take the final election of the members of our future panel. The number of sessions will depend on the homogeneity between candidates or of the amplitude of the differences generate after each test between subjects. Only when a reasonable security exits on the possibility of choosing the best individuals, the selective process should be stopped.

Following to the selection, the training period will start with two different phases: global training and specific (on meat) training. Globally, this period should include a general training in the basic sensory procedure including how to manage the cabin facilities and the use of scales, a rational program to increase the sensory abilities of the panel members including the knowledge of their thresholds (stimulus perception and recognition) and a teaching program to get a deep and compressive knowledge of the product.

Also, it is necessary to get a perfect understanding of the global sensory rules (be on time for the start of the session, do not smoke or have strong flavoured food or drink before a test, no need to hurry, behaviour inside the cabins, etc.). It is basic to make the panel members understand that they form a group and that the main objective is not that all the assessors have the same opinion (if that was the case, only one person would be enough to form a taste panel), but each member must be coherent with its own criteria and stable opinions in the time. In any case, it is necessary that the panel have a unique terminology and a common vocabulary is then a basic objective (ISO-5492). It is necessary to make some statistical analysis to verify the individual and group repeatability (Guerrero and Guardia, 1998).

Other important point is the specific sampling procedures and the preparation of the samples prior to the analysis, including sampling, cooking and serving. The muscle should be, at least, the *longissimus dorsi*. Samples should be cooked entirely with subcutaneous fat included, in a grill preheated to 200 °C, without any condiment and wrapped in domestic aluminium foil and adequately identified. After cooking (until a fixed internal temperature: recommended 70 °C but with variations between 65 °C in French style and 80 °C in countries with an overcooked appreciate taste), meat should be sliced perpendicularly to the main muscle axe in 2 cm-thick chops. Each assessor, within the same dish, should taste the same or the most possible similar pieces and the same piece of the slice considering: lateral (L), central (C) or medial (M), or just lateral or medial in light carcasses (Figure 2).

Samples must be kept hot at every time (50-60 °C depending of final temperature considered) and hot dishes must be provided during the tasting. In each taste session (1 hour) not more than 12 samples must be tasted, and not more than two, or exceptionally three, sessions by day, with half an hour of break between sessions. Samples must be presented in appropriate groups, in each dish, following the

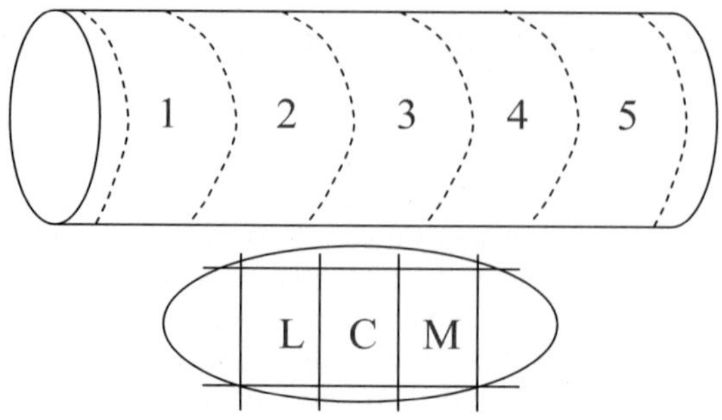

Figure 2. Slice (1-4) and sample [lateral (L) - central (C) - medial (M)] distribution within a slice for each consumer (lamb).

model of equilibrate comparison (Cohram and Cox, 1978) indispensable in the sensory routine. Each dish should be served together to the panel disposition, or samples be served consequently, but tested in different order by each panellist to avoid first-over and carry-over effects (MacFie *et al.*, 1989). Between dishes, a brake of approximately 3 minutes, without interruptions, should be considered. If the procedure is adequate, the results of the sensory analysis will be comparable between countries in the objective descriptors of the profile, not in the hedonic ones, as was pointed out by Sañudo *et al.* (1998b) (Table 3). Another important point is the ageing period before analysis. We should consider that long ageing periods tend to homogenate sensory meat quality, although long periods in carcass, before vacuum packaging, affect negatively the colour (unpublished) (Table 4).

Table 3. Sensory attributes of 4 lamb types evaluated by both Spanish and British panels.

	Spanish Merino concentrate fed		Rasa Aragonesa, concentrate fed		Welsh Mountain, reared on grass		Suffolk x Mule reared on grass	
Odour intensity								
British panel	2.8 [b]	0.20	2.71 [b]	0.16	4.0 [a]	0.21	3.61 [a]	0.21
Spanish panel	57.9 [b]	2.14	58.51 [b]	1.34	67.34 [a]	1.99	-	-
Tenderness								
British panel	6.2 [a]	0.14	5.3 [b]	0.14	6.5 [a]	0.15	6.05 [a]	0.14
Spanish panel	72.5 [a]	2.06	58.5 [b]	1.64	69.3 [a]	2.06	-	-
Juiciness								
British panel	5.2 [a]	0.12	5.1 [ab]	0.12	5.1 [ab]	0.14	4.8 [b]	0.13
Spanish panel	64.1 [a]	2.01	61.7 [a]	1.39	45.11 [b]	2.23	-	-
Flavour intensity								
British panel	3.0 [b]	0.17	3.2 [b]	0.16	4.4 [a]	0.19	3.9	0.18
Spanish panel	63.6 [b]	1.94	58.2	1.23	73.7	1.82	-	-
Flavour quality								
British panel	3.3 [b]	0.17	3.3 [b]	0.15	4.5 [a]	0.20	4.0	0.20
Spanish panel	57.8 [a]	1.72	59.4 [a]	1.13	33.1 [b]	1.85	-	-
Overall acceptability								
British panel	3.5 [bc]	0.16	3.4 [c]	0.15	4.7 [a]	0.20	4.1 [b]	0.19
Spanish panel	59.5 [a]	1.92	57.8 [a]	1.34	32.0[b]	2.01	-	-

N= 88 for the Spanish panel; n= 80 for the British panel.

The British panel used an 8 points scale and the Spanish one of 100 points.

[a,b] Different superscripts within rows indicate significant differences between lamb types.

Table 4. Ageing effect, in carcass + 3 weeks in vacuum, on light lamb meat quality.

	3 days without vacuum	1 day + vacuum	4 days + vacuum	8 days + vacuum
L*	43.1 [c]	41.1 [b]	42.1 [b]	34.7 [a]
a*	15.8 [a]	17.9 [b]	15.3 [a]	20.8 [c]
b*	10.2 [a]	11.7 [a]	10.4 [a]	18.7 [b]
C20% (N/cm^2)	7.32 [b]	5.05 [b]	4.71 [b]	5.67 [b]
WB (kg)	2.9 [b]	1.8 [ab]	1.9 [ab]	1.7 [ab]
Tenderness	71.1 [a]	81.8 [b]	82.0 [b]	81.7 [b]
Lamb flavour intensity	68.7	63.3	66.8	65.9
Liver flavour intensity	39.4 [a]	49.9 [b]	52.5 [b]	50.9 [b]
Consumers acceptability	55.5 [a]	71.4 [b]	67.1 [b]	67.8 [b]

[a-c] Different superscripts within rows indicate significant differences between lamb types.

The profile, group of attributes that describe the product, must be perfectly understood by all the panel members and a consensus about it must be necessarily obtained. The different attributes must be in a logical position, following the common order of their perception. A large profile for lamb has been described by Jeremiah (1988). The most use and recommendable for cooked lamb meat, except if a visual test is required (self-life, or consumer choice approximation), are the following:

- *Odour*. evaluated immediately after unwrapping the sample.
  - Livery. odour similar to heated lamb liver.
  - Blood, odour similar to fresh blood.
  - Milky, odour similar to ewe fresh milk.
  - Mutton or lamb, similar to the specific ovine odour.
  - Woolly, odour similar to the fresh wool after shearing.
  - Off odour, atypical odour often associated with deterioration or transformation of the product.
- *Texture*. evaluated at the firs bit and during the chewing process.
  - Tenderness, multidimensional characteristic related with the facility to deform the product and to be divided in particles by the molar teeth.
  - Juiciness, perception of moisture content of the product released during the chewing process.
  - Chewiness, mechanical attribute related to the amount of work required to masticate the product into a state ready for swallowing.
  - Fibrosity, geometrical attribute relating to the perception, during chewing, of long particles oriented in the same direction.
  - Fattiness, perception of fat on the surface or in the body of the product.
- *Flavour*. complex combination of the olfactory, gustatory and trigeminal sensations perceived during tasting, evaluated during the chewing.
  - Livery, flavour similar to heated lamb liver.
  - Fatty, flavour similar to lamb fat.
  - Mutton or lamb, similar to the specific ovine taste.
  - Acidic, primary taste produced by dilute aqueous solutions of mist acid substances (e.g., citric, tartaric acids).
  - Milky, flavour similar to ewe fresh milk.
  - Off flavour, atypical flavour often associated with deterioration or transformation of the product.
- *After taste*. evaluated after deglutition.
  - Sour, gustatory complex sensation, generally due to presence of organic acids.

The sensory profile should be presented to the panel using non-structured 100 mm line scales, in which 0 represents the smallest intensity or null perception of the respective attribute (not odour or flavour, very tough meat, very dry and so on) and 10, or 100, the highest (very high odour or flavour intensity, very tender, very juicy meat and so on).

The test should be run in standard cabins under red light, bread and still water must be provided and the mouth should be cleaned between samples. An appropriate statistical analysis, considering the effects should be done for a correct result discussion. Session and panellist must be considered as fix effects.

Consumer analysis

A consumer test is a hedonic analysis in which assessors express their own opinion about a product, expressing whether they like it or not, whether they prefer it to another product or not (Campo, 2005). One of the main aspects is the selection of consumers, who should be representative of the population we want to study. We do not have to select them according to their aptitudes to detect organoleptic differences in the meat since the preference of a consumer towards any product can be influenced by physical, social, economical or environmental characteristics but, especially, by their previous experience about the product. In order to perform a correct selection of consumers we should take into account:

- *Gender.* Ideal values should match the initial population, normally 50% for men and women.
- *Age.* Consumers should represent the age distribution, considering that certain populations have more elderly people and others a higher birth rate.
- *Geographical location.* If we want results that apply for a wide region, maybe we need to consider two or more paces of study if they have different consumption habits.
- *Qualifications.* Depending on the aim of the study, the qualifications of the consumer may influence the result of the test due to the lack or not of previous experience about the product.
- *Income.* Higher economical incomes vary the consumption of fresh meat and, therefore, the preferences of consumers.
- *Confession.* We should not perform a pork test on a Muslim population.

A consumer test can be performed in different locations (Meilgaard *et al.*, 1991):

- *Laboratory.* Taking consumes into a research area have several advantages:
  - Environmental conditions (colour masking, refrigeration, individual booths, etc), can be controlled, as well as aspects about preparation and presentation of samples.
  - If part of the population belongs to the working area, it is easy and fast to attend the test.
  - 100% recovery of questionnaires.
  - Consumers work more focused.

  However, it has several disadvantages:
  - If consumers are away from the location, they need to arrange transport to attend the test.
  - Certain consumers may have previous knowledge of the parameters to study on the analysis.
  - The conditions of the test are not the real ones.
- *Home tests.* Doing a consumer test in somebody's home can be positive due to:
  - The cooking of the samples is done under real conditions.
  - More information about the family can be obtained.
  - The information can reflect the social conditions.

  Nevertheless:
  - We can not control the conditions in which the test is being preformed.
  - There may be a dominant consume within the family who influences the opinion of the rest.

- We need a higher amount of sample and a longer period of time to perform the test.
- Any doubt by the consumer about the test can not be answered on time.

There is another option of performing consumer tests in central locations (Supermarkets, Malls). However, it is more difficult to do due to the need of cooking or refrigeration of fresh meat, which requires appropriate facilities. A sample of different consumer test in lamb is shown in Table 5.

There are two types of methods performing consumer tests:

- *Qualitative*. These methods assess the preference of consumers through dialogues. The most popular tests are *focus groups*, where different groups of consumers (8-12 people) meet for 1-2 hours with a team leader to talk about a set subject, or *individual interviews*, where the dialogue exist between the leader and each consumer individually. The last one is recommended with sensitive subjects or with timid people unable to express themselves in public.
- *Quantitative*. With these methods, the preference of consumers towards a product or a specific characteristic can be assessed.

Preference test. The consumer chooses a product in comparison with another, but it does not give any information whether he likes it or not. Samples can be presented in:

- Couples. A sample must be chosen out of two (A-B).
- Ranking. A preference order has to be expressed (A-B-C).
- Multiple couples (all). It is a couples test in a balance series of several samples (A-B, A-C, B-C).
- Multiple couples (selection). It is a couples test with a control sample (X-A, X-B, X-C).

Acceptability test. The consumer indicates how much he likes or dislikes a sample, within a hedonic scale. The number of possible answers, even or odd, implies that the consumer has to decide positively or negatively (even) (Table 6) or can be neutral (odd).

Table 5. Different consumer tests performed with lamb.

	n	Location	Sample	Cooking
Dransfield *et al.*, 2000	216 (6 countries)	Home	Leg	Oven
Martínez-Cerezo *et al.*, 2005c	265	Laboratory	ST, SM, GB	Grill 200 °C
Font i Furnols *et al.*, 2006	600 (3 countries)	Laboratory	LD	Grill

ST: Semitendinosus; SM: Semimembranosus; GB: Gluteo biceps; LD: Longissimus dorsi.

Table 6. Preferences of German, Spanish and British consumers (%) depending on the type of lamb evaluated, Uruguayan or local aged for 7 or 20 days in each country (Font i Furnols et al., *2006).*

	n	Like extremely Like very much Like moderately	Like slightly Dislike slightly	Dislike extremely Dislike very much Dislike moderately
UY light	200	64.0 ^{b}	33.0 ^{a}	3.0
UY heavy		76.5 ^{ab}	21.0 ^{ab}	2.5
DE 7d		73.5 ^{ab}	25.0 ^{a}	1.5
DE 20d		83.0 ^{a}	16.0 ^{b}	1.0
UY light	200	59.8	37.0	3.2 ^{b}
UY heavy		53.5	36.9	9.6 ^{a}
ES 7d		60.3	33.0	6.8 ^{ab}
ES 20d		54.7	33.5	11.8 ^{a}
UY light	192	51.0 ^{b}	34.4 ^{a}	14.6 ^{a}
UY heavy		80.8 ^{a}	15.6 ^{b}	3.7 ^{b}
UK 7d		78.1 ^{a}	19.8 ^{b}	2.0 ^{b}
UK 20d		79.2 ^{a}	17.7 ^{b}	3.1 ^{b}

UY: Uruguay (grazing); Light: 11.4 kg carcass weight, 3-4 months old; Heavy: 19.4 kg carcass weight, 12-13 months old; DE: Germany (finished on pasture); ES: Spain (reared on concentrates); UK: United Kingdom (Grazing + supplementation with concentrates).

^{a,b} Different superscripts within columns and attributes indicate significant differences ($P<0.05$).

Chemical analysis

Meat from ruminants is mainly consumed cooked; this process modifies the composition of the raw material. The amount of minerals in cooked meat increases with cooking because it increases the amount of dry matter, since cooking dehydrates the muscle (Table 7). However, the amount of vitamins remains almost the same in raw than in cooked meat even with that increment in dry matter, meaning that there is a loss during cooking due to the temperature or to the losses in juice. Chemical analysis must be done in a representative sample of the muscle to be analysed, free of external fat and connective tissues, although it is often used to predict the status of the whole joint or even the whole carcass. There is however variation within muscle, between muscles and between animals. Meat should be minced with an electrical mixer avoiding water loses and kept at 4 °C sealed in a full container for less than 24 hours for analysing dry matter. Then, the minced sample can be vacuum packaged and freeze (-24 °C) to perform the rest of analysis up to three months later. Dry matter is recommended to be analysed again after thawing.

The techniques to be used for determining the chemical composition of meat are:

- Moisture (ISO R-1442). Based on the difference of weight before and after drying meat at 100 °C.
- Fat (ISO R-1443). The procedure weights the residue left in the evaporation of hexane or petroleum ether, used in the extraction of the fat.
- Protein (ISO R-937). Protein is calculated from the content of nitrogen in the sample x 6.25.
- Ashes (ISO R-936). The sample is kept at 550 °C with the addition of magnesium acetate.
- Collagen (Bonnet and Kopp, 1984). The analysis detects the amount of hydroxiproline, amino acid that composes 12.5% of connective tissue. The solubility of the collagen can be assessed after different combinations of cooking temperatures and lengths of time, from the difference between the hydroxiproline in the intact sample and the one extracted from the sample (Hill, 1966).

Table 7. Chemical composition in raw and cooked meat from lamb and goat.

	Lamb [1]		Goat [2]
	Raw	Cooked	Cooked
Moisture (%)	76.5	66.5	58.4
Protein (g/100g)	20.4	28.7	26.1
Fat (g/100g)	2.7	4.2	13.2
Energy (kcal/100g)	105.0	153.0	229.5
Fe (mg/100g)	1.70	2.20	2.20
Zn (mg/100g)	2.75	3.59	4.40
Cu (mg/100g)	0.17	0.18	1
Vit. B_1 (mg/100g)	0.18	0.14	0.24
Vit. B_2 (mg/100g)	0.19	0.26	0.27
Vit. B_3 (mg/100g)	5.76	5.19	3.48
Vit. B_5 (mg/100g)	0.31	0.33	0.38
Vit. B_6 (mg/100g)	0.34	0.26	0.21
Vit. B_{12} (μg/100g)	3.08	3.60	1.06

[1] Badiani *et al.*, 1997 (90 days old, 17.7 kg carcass weight).

[2] Johnson *et al.*, 1995 (6-8 months old, 10.7-13.8 kg carcass weight).

- Fatty acids (Folch *et al.*, 1957; Bligh and Dyer, 1959). The extraction of fatty acids is based on their solubility properties, i.e., in choloroform. After the extraction, a methylation must be used in order to analyse them by gas chromatography. If isomers are of interest, acid methylation should be avoided. In order to quantify (mg/100 g of sample), and internal standard should be used, such as heneicosanoic acid (C21:0).

Age together with weight decreases total collagen content in the muscle (Table 8), probably because it increases other components, such as intramuscular fat. However, the amount of insoluble collagen increases due to the increment of non-soluble cross links, which reduces the solubility of the collagen, contributing to a tougher meat. Goat's meat has a higher content of total collagen and lower solubility than lamb's meat, especially when comparing lambs of 6 months old versus goats of 11 months old.

Table 8. Total (Ht) and insoluble (Hi) hydroxyproline and collagen solubility (%S) in lamb from lamb breeds (slaughtered at different weights) and goat.

Breed	Rasa Aragonesa [1]		Churra [1]		Goat [2]
Live weight (kg)	10-12	30-32	10-12	30-32	
Ht	368.45	313.78	421.44	387.87	565.00
Hi	156.96	201.45	190.08	251.28	-
% S	54.93	34.01	55.23	32.34	13.40

[1] Martínez-Cerezo *et al.*, 2005a.

[2] Kannan *et al.*, 2006 (11 months old).

The composition of fatty acids in the intramuscular fat depends on several factors such as age, weight, diet, breed and specie. At the same slaughter weight, younger animals show less intramuscular fat than older ones (Table 9), even when raised on a more energetic diet (concentrate *vs.* grazing). Goats fatten less than lambs, and in order to reach a similar 3% of intramuscular fat, goats have to be quite older. Even then, its carcass weight is much lower than in lambs.

Animals fed under intensive conditions with a cereal-based diet show a higher composition in *n*-6 fatty acids, especially linoleic acid (C18:2 *n*-6), characteristic of concentrates. However, those fed on grass show higher content in n-3 fatty acids, especially α-linolenic acid (C18:3 *n*-3).

Lipid oxidation of meat will depend on fatty acid composition, storage conditions and the amount of antioxidant present the muscle. Grass is rich in natural antioxidants and for this reason, without supplementation, the shelf life of meat from grass fed animals is longer than from concentrate fed ones. A supplementation of 250 mg of vitamin E per kg of feed in intensively fed systems is enough to avoid lipid oxidation increasing the shelf life of the product (Figure 3).

Table 9. Fatty acid composition of intramuscular fat of longissimus dorsi muscle.

	Rasa Aragonesa [1]	Welsh Mountain [2]	Corriedale [3]	Goat [4]
IM (%)	1.99	3.69	3.05	3.00
14:0	3.51	2.43	3.60	1.50
16:0	22.04	22.96	24.73	22.20
18:0	12.87	21.12	16.62	15.10
18:1	36.78	39.46	35.81	44.50
18:2 *n*-6	9.10	2.41	6.01	3.10
18:3 *n*-3	0.60	1.81	3.37	0.24
20:4 *n*-6	0.23	0.96	1.94	1.30
20:5 *n*-3	0.69	0.73	1.14	0.15
22:6 *n*-3	0.29	0.24	0.31	0.05

[1] Sañudo *et al.*, 2000a (80 days old; 10-11 kg carcass weight, concentrate fed).
[2] Sañudo *et al.*, 2000a (5-6 months old; 11 kg carcass weight, grass fed).
[3] Díaz *et al.*, 2005 (3-4 months old, 11.1 kg carcass weight, grass fed).
[4] Bas *et al.*, 2005 (Moroccan breed, 12-14 months old, 7.9 kg carcass weight).

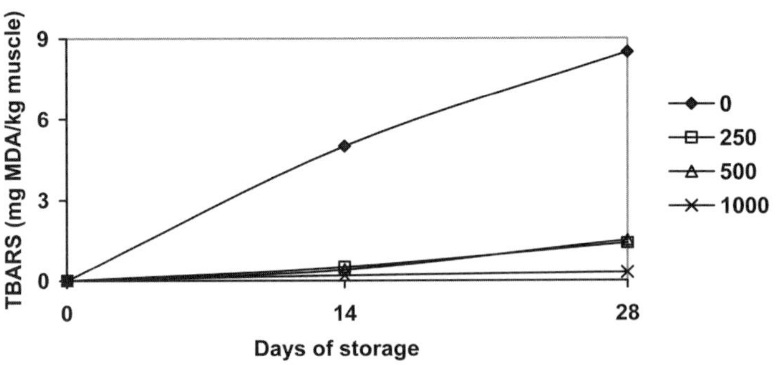

Figure 3. Effect of supplementation with Vitamine E in the diet of lambs (mg/kg feed) on the oxidation of m. longissimus dorsi throughout storage (Lauzurica et al.*, 2005).*

Volatile analysis can be performed in different tissues. The analysis has different stages: extraction of components, desorbtion, identification and quantification. The extraction implies that the volatiles are swept from a fix weight of the mince sample, normally after cooking, placed on a sealed flask, by an inert gas. The most used are helium and nitrogen. Afterwards, the compounds must be trapped on a column filled with either Tenax® or another resin (i.e. Lichrolut®). Desorbtion can be performed directly on the headspace of the gas chromatograph or by a solvent that will be injected afterwards. In any case, a gas chromatograph will be used to separate compounds on low bleed fused silica capillary column, fitted with a mass spectrometer.

References

Alcalde, M.J., C. Sañudo, J.C. Osorio, J.L. Olleta and I. Sierra, 1999. Evaluación de la calidad de la canal y de la carne en canales ovinas ligeras del tipo comercial 'Ternasco'. Información Técnica Económica Agraria, Volume 95A, 49-64.

Alexandrova, N., V. Banskalieva, A. Angelov, I. Ivanov, S. Laleva and P. Slavova, 1996. Meat quality characteristics and fatty acid composition of triacylglicerols in out-of-season born lambs. 42nd ICoMST, 204-205.

Badiani, A., N. Nanni, P.P. Gatta, F. Bitossi, B. Tolomelli and M. Manfredini, 1997. Nutrient content and retention in selected roasted cuts from 3-month old ram lambs. Food Chemistry, Volume 61, 89-100.

Bas, P., E. Dahbi, A. El Aich, P. Morand-Fehr and A. Araba, 2005. Effect of feeding on fatty acid composition of muscles and adipose tissues in young goats raised in the Argan tree forest of Morocco. Meat Science, Volume 71, 317-326.

Berge, P., A. Sánchez, C. Sañudo, G. Thorkelsson, C. Stamataris, E. Piasentier and A. Fisher, 2000. Variation in muscle composition between different commercial lamb types and its relationship with meat texture. 46th ICoMST, 106-107.

Beriain, M., A. Horcada, A. Purroy, G. Lizado, J. Chasco and J. Mendizábal, 2000. Characteristics of Lacha and Rasa Aragonesa lambs slaughtered at three live weights. Journal of Animal Science, Volume 78, 3070-3077.

Bligh, E.G. and Dyer, W.J., 1959. A rapid method for total lipid extraction and purification. Canadian Journal of Biochemistry and Physiology, Volume 37, 911–917.

Boccard, R., L. Buchter, E. Casteels, E. Cosentino, E. Dransfield, D.E. Hood, R.L. Joseph, D.B. Magdougal, D.N. Rhodes, I. Schön, B.J. Tinbergen and C. Touraille, 1981. Procedures for measuring meat characteristics in beef production experiments. Report of a working group in the Commission of the European Communities (CEC) Beef Production Research Programme. Livestock Production Science, Volume 8, 385-397.

Bonnet, M. and J. Kopp, 1984. Dosage du collagène dans les tissues conjonctifs, la viande e tales produits carnés. Cahiers Technologiques INRA, Volume 5, 19-30.

Bouton, P.E., P.V. Harris and W.R. Shorthose, 1971. Effect of ultimate pH upon the water-holding capacity and tederness of mutton. Journal of Food Science, Volume 36, 435-439.

Bouton, P.E. and P.V. Harris, 1972. The effects of cooking temperature and time on some mechanical properties of meat. Journal of Food Science, Volume 37,: 140-144.

Bray, A.R., S.R. Young and G.H. Scales, 1994. Variation in the pH of lamb meat within and between sheep breeds. Proceedings of the New Zealand Society of Animal Production, Volume 54, 201-203.

Campo, M.M., 2005. Consumidores. In: Cañeque, V. and C. Sañudo (eds.), Estandarización de las metodologías para evaluar la calidad del producto (animal vivo, canal, carne y grasa) en los rumiantes. Monografías INIA: Serie Ganadera, Volume 3, 409-422.

Cañeque, V., C. Pérez, S. Velasco, M.T. Díaz, S. Lauzurica, I. Alvarez, F. Ruiz de Huidobro, E. Onega and J. De la Fuente, 2004. Carcass and meat quality of light lambs using principal component analysis. Meat Science, Volume 67, 595-605.

Commission Internationale de l'Eclariage, 1986. CIE Colorimetry, 2nd ed., Publication CIE 15.2, Vienna, Austria.

Cochran, W.G. and G.M. Cox, 1978. Diseños Experimentales. Ed. Trillos, México.

Devendra, C., 2000. Challeges for research and development of goats. Proceedings of the 7th International Conference on Goats, Tours, France, Volume 2, 887-889.

Devine, C.E. and A.E. Graafhuis, 1995. The basal toughness of unaged lamb. Meat Science, Volume 39, 285-291.

Devine, C.E., A.E. Graafhuis, P.D. Muir and B.B. Chrystall, 1993. The effect of growth rate and ultimate pH on meat quality of lambs. Meat Science, Volume 35, 63-77.

Díaz, M.T., S. Velasco, V. Cañeque, S. Lauzurica, F. Ruíz de Huidobro, C. Pérez, J. González and C. Manzanares, 2002. Use of concentrate or pasture for fattening lambs and its effect on carcass and meat quality. Small Ruminant Research, Volume 43, 257-268.

Díaz, M.T., S. Velasco, C. Pérez, S. Lauzurica, F. Huidobro and V. Cañeque, 2003. Physico-chemical characteristics of carcass and meat Manchego-breed suckling lambs slaughtered at different weights. Meat Science, Volume 65, 1085-1093.

Díaz, M.T., I. Álvarez, J. De la Fuente, C. Sañudo, M.M. Campo, M.A. Oliver, M. Font i Furnols, F. Montossi, R. San Julián, G.R. Nute and V. Cañeque, 2005. Fatty acid composition of meat from typical lamb production systems of Spain, United Kingdom, Germany and Uruguay. Meat Science, Volume 71, 256-263.

Dransfield, E., G.R. Nute, B.W. Hogg and B.R. Walters, 1990. Carcass and eating quality of ram, castrated ram and ewe lambs. Animal Production, Volume 50, 291-299.

Dransfield, E., J.F. Martin, A. Fisher, G. Nute, C. Zygyiannis, D. Stamataris, G. Thorkelsson, T. Valdimardottir, E. Piasentier, C. Mils, C. Sañudo and M. Alfonso, 2000. Home placement testing of amb conducted in six countries. Journal of Sensory Studies, Volume 15, 421-436.

Faustman, C. 1994. Postmortem changes in muscle foods. In: Kinsman, D.M., A.W. Kotula and B.C. Breidenstein (eds.), Muscle foods: Meat, poultry and seafood technology. Ed. Chapman & Hall, New York and London.

Field, R.A., J.C. Williams, C.L. Ferrel, J.D. Crouse and J.E. Kunsman, 1978. Dietary alteration of palatability and fatty acids in meat from light and heavy weight ram lambs. Journal of Food Science, Volume 47, 858-864.

FAOSTAT, 2001. http://faostat.fao.org/site/336/default.aspx.

Fogarty, N.M., D.L. Hopkins and R. Vande Ven, 2002. Lamb production from diverse genotypes. 2. Carcass characteristics. Animal Science, Volume 70, 147-156.

Folch, J., Lees, M. and Sloane-Stanley, G.H., 1957. A simple method for the isolation and purification of total lipids from animal tissues.. Journal of Biological Chemistry, Volume 226, 497–509.

Font i Furnols, M., R. San Julián, L. Guerrero, C. Sañudo, M.M. Campo, J.L. Olleta, M.A. Oliver, V. Cañeque, I. Álvarez, M.T. Díaz, W. Branscheid, M. Wicke, G.R. Nute and F. Montossi, 2006. Acceptability of lamb meat from different producing systems and ageing time to German, Spanish and British consumers. Meat Science, Volume 72, 545-554.

Gault, N.F.S., 1985. The relationship between water holding capacity and cooked meat tenderness in some beef muscles as influenced by acidic conditions below the ultimate pH. Meat Science, Volume 15, 15-30.

Guerrero, L., 2005. Panel entrenado. In: Metodología para el estudio de la calidad de la canal y de la carne en ruminates. Monografías INIA: Ganadera, Volume 1, 397-408.

Guerrero, L. and M.D. Guardia, 1998. Evaluación de la fiabilidad de un panel de cata. III Jornadas de Análisis Sensorial, Valdediós, Asturias, España.

Grau, R. and R. Hamm, 1953. Eine einfache Methode zur Bestimmurg der Wassesbindung mi Muskel. Die Noturwissenschaften, Volume 40, 29-30.

Griffin, C.L., M.W. Orcutt, R.R. Riley, G.C. Smith, J.W. Savell and M. Shelton, 1992. Evaluation of palatability of lamb, mutton and chevron by sensory panels of various cultural backgrounds. Small Ruminant Research, Volume 8, 67-74.

Hamm, R., 1960. Biochemistry of meat hydratation. In: Chichester, C. O., E. M. Mrak and G. F. Stewart (eds.), Advances Food Research, Academic Press, New York and London, Volume 10, 355.

Hill, F., 1966. The solubility of intramuscular collagen in meat animals of various ages. Journal of Food Science, Volume 31, 161-166.

Honikel, K.O., 1991. Assessment of meat quality. In: Fiems, L.O. and B.G. Cottyn (eds.), Animal biotechnology and the quality of meat production. Elsevier, Amsterdam, The Netherlands, 107-125.

Honikel, K.O., 1997. Reference methods supported by OECD and their use in Mediterranean meat products. Food Chemistry, Volume 5, 573-582.

Honikel, K.O., 1998. Reference methods for assessment of physical characteristics of meat. Meat Science, Volume 49, 447-457.

Hopkins, D.L., D.G. Hall, H.A. Channon and P.J. Holst, 2001. Meat quality of mixed sex lambs grazing pasture and supplemented with, roughage, oats or oats and sunflower meal. Meat Science, Volume 59, 277-283.

Horcada, A., 1996. Calidad de la canal de los corderos de las razas Lacha y Rasa Aragonesa. Tesis Doctoral. Universidad Pública de Navarra, Pamplona, España, 252 pp.

International Standard ISO 1442, 1973. Meat and meat products. Determination of moisture.

International Standard ISO 1443, 1973. Meat and meat products. Determination of total fat content.

International Standard ISO 936, 1978. Meat and meat products. Determination of ash.

International Standard ISO 937, 1978. Meat and meat products. Determination of nitrogen content.

Jaime, I., J.A. Beltrán, P. Ceña, P. López-Lorenzo and P. Roncalés, 1992. Tenderisation of lamb meat: effect of rapid postmortem temperature drop on muscle conditioning and aging. Meat Science, Volume 32, 357-366.

Jansen, M.L., 2001. Determination of meat pH-temperature relationship using ISFET and glass electrode instruments. Meat Science, Volume 58, 145-150.

Jeremiah, L.E., 1982. Factors affecting consumption, selection and acceptability of meat purchases. Journal of Consumer studies and Home Economics, Volume 6, 137-154.

Jeremiah, L.E., 1988. A combination of flavour and texture profiles for lamb leg roast from three different geographical sources. Canadian Institute of Food Science and Technology Journal, Volume 5, 471-476.

Jeremiah, L.E., 2000. The effects of chronological age, slaughter weight and gender on lamb: a review. Lacombe Research Centre, Technical Bulletin 2000-1E, 1-19.

Johnson, D.D., J.S. Eastridge, D.R. Neubaueer and C.H. McGowan, 1995. Effect of sex class on nutrient content of meat from young goat. Journal of Animal Science, Volume 73, 296-301.

Kannan, G., K.M. Gadiyaram, S. Galipalli, A. Carmichael, B. Kouakou, T.D. Pringle, K.W. McMillin and S. Gelaye, 2006. Meat quality in goats as influenced by dietary protein and energy levels, and postmortem aging. Small Ruminant Research, Volume 61, 45-52.

Koohmaraie, M., S.D. Shackelford, T.L. Wheeler, S.M. Lonergan and M.E. Doumit, 1995. A muscle hypertrophy condition in lamb (callipyge): characterization of effects on muscle growth and meat quality traits. Journal of Animal Science, Volume 73, 3596-3607.

Lanza, M., M. Bella, A. Priolo, D. Barballo, V. Galofaro, C. Landi and P. Pennisí, 2006. Lamb meat quality as affected by a natural or artificial milk feeding regime. Meat Science, Volume 73, 313-318.

Lauzurica, S., J. De la Fuente, M.T. Díaz, I. Álvarez, C. Pérez and V. Cañeque, 2005. Effect of dietary supplementation of vitamin E on characteristics of lamb meat packed under modified atmosphere. Meat Science, Volume 70, 639-646.

Lawrie, R.A., 1985. Meat Science. Pergamonn Press, Oxford, UK.

Lawrie, R.A., 1998. Glucólisis post mortem. En: Acribia, S.A. (ed.), Ciencia de la carne. Zaragoza, España. 367 pp.

Lawrie, R.A., 2002. The eating quality of meat. In: Meat Science, 5th ed., Pergamon Press., New York, pp. 173-176, 184-188.

Lebbie, S., 2000. Organisation of goat research: the African perspectives. In: Proceedings of the Seventh International Conference on Goats, Tours, France, Volume 2, 890-891.

Lepetit, J., 1991. Theoretical strain ranges in raw meat. Meat Science, Volume 29, 271-283.

Lepetit, J. and J. Culioli, 1994. Mechanical properties of Meat. Meat Science, Volume 36, 203-237.

MacFie, H.J.H., Bratchell, N., Greenhoff, K. and Vallis, L.V. (1989). Designs to balance the effect of order of presentation and first order carry-over effects in hall tests. Journal of Sensory Studies, Volume 4, 129-148.

Martínez-Cerezo, S., C. Sañudo, B. Panea, I. Medel, R. Delfa, I. Sierra, J.A. Beltrán, R. Cepero and J.L. Olleta, 2005a. Breed, slaughter weight and ageing time effects on physico-chemical characteristics of lamb meat. Meat Science, Volume 69, 352-333.

Martínez Cerezo, S., C. Sañudo, I. Medel and J.L. Olleta, 2005b. Breed, slaughter weight and ageing time effects on sensory characteristics of lamb. Meat Science, Volume 69, 571-578.

Martínez-Cerezo, S., C. Sañudo, B. Panea and J.L. Olleta, 2005c. Breed, slaughter weight and ageing time effects on consumer appraisal of three muscles of lamb. Meat Science, Volume 69, 797-805.

Meilgaard, M., G.V. Civille and B.T. Carr, 1991. Sensory evaluation techniques. CRC Press, Boca Raton, FL, USA, 354 pp.

Monin, G., 1989. Facteurs biologiques des qualites de la viande. Colloque INRA-ENSA de Croissance bovins et qualité viande, Rennes, France, 177.

Morand-Fehr, P. and S.H.B. Lebbie, 2004. Proposals for improving the research efficiency in goats. Small Ruminant Research, Volume 51, 145-153.

Okeudo, N.J., B.W. Moss and M.B. Chestnutt, 1994. Effect of feeding a milk diet or concentrate plus hay diet on carcass and meat quality of lamb. 40th ICoMST, S-IVA.38.

Pearson, A.M. and R.B. Young, 1989. Postmortem conversion of muscle to meat. In: Muscle and Meat Biochemistry. Academic Press, San Diego, USA, 391-444.

Priolo, A., D. Micol, J. Agabriel, S. Prache and E. Dransfield, 2002. Effect of grass or concentrate feeding systems on lamb carcass and meat quality. Meat Science, Volume 62, 179-185.

Purslow, P., 2005. Intramuscular connective tissue and its role in meat quality. Meat Science, Volume 70, 435-447.

Renerre, M., 1986. Influence des facteurs biologiques et tecnologiques sur le couleur de la viande bovine. Bulletin Technique Centre de Recherches Zootechniques et Vétérinaires de Theix-INRA, Volume 65, 41-45.

Rousset-Akrim, S., O.A. Young and J.L. Berdagué, 1997. Diet effects and growth effects in panel assessment of sheep meat odour and flavour. Meat Science, Volume 45, 169-181.

Russo, C., G. Preziuso and P. Verità, 2003. EU carcass classification system: carcass and meat quality in light lambs. Meat Science, Volume 64, 411-416.

Santos-Silva, J., I.A. Mendes and R.J.B. Bessa, 2002. The effect of genotype, feeding system and slaughter weight on the quality of light lambs. 1. Growth, carcass composition and meat quality. Livestock Production Science, Volume 76, 17-25.

Santos-Silva, J., R.J.B. Bessa and I.A. Mendes, 2003. The effect of supplementation with expanded sunflower seed on carcass and meat quality of lambs raised on pasture. Meat Science, Volume 65, 1301-1308.

Sañudo, C., M.P. Santolaria, G. María, M. Osorio and I. Sierra, 1996. Influence of carcass weight on instrumental and sensory lamb meat quality in intensive production systems. Meat Science, Volume 42, 195-202.

Sañudo, C., M.M. Campo, I. Sierra, G. María, J.L. Olleta and P. Santolaria, 1997. Breed effect on carcass and meat quality of suckling lambs. Meat Science, Volume 46, 357-365.

Sañudo, C., A. Sánchez and M. Alfonso, 1998a. Small ruminant production systems and factors affecting lamb meat quality. Meat Science, Volume 49, S29-S64.

Sañudo, C., I. Sierra, J.L. Olleta, L. Martín, M.M. Campo, P. Santolaria, J.D. Word and G.R. Nute, 1998b. Influence of weaning on carcass quality, fatty acid composition and meat quality in intensive lamb production systems. Animal Science, Volume 66, 175-187.

Sañudo, C., E. Enser, M.M. Campo, G.R. Nute, G. María, I. Sierra and J.D. Word, 2000a. Fatty acid composition and sensory characteristics of lamb carcasses from Britain and Spain. Meat Science, Volume 54, 339-346.

Sañudo, C., M. Alfonso, A. Sánchez, R. Delfa and A. Teixeira, 2000b. Carcass and meat quality in light lambs from different fat classes in the EU carcass classification system. Meat Science, Volume 56, 89-94.

Sañudo, C., M. Alfonso, A. Sánchez, P. Berge, E. Dransfield, D. Zygoyiannis, C. Stamataris, G. Thorkelsson, T. Vladimarsdottir, E. Piasentier, C. Mills, G.R. Nute and A.V. Fisher, 2003. Meat texture of lambs from different European production systems. Australian Journal of Agricultural Research, Volume 54, 551-560.

Sañudo, C., J.L. Olleta, M.M. Campo, B. Panea and E. Rota, 2005. Propuesta de Muestreo. In: Estandarización de las metodologías para evaluar la calidad del producto (animal vivo, canal, carne y grasa) en los rumiantes. Monografías INIA: Serie Ganadera, Volume 3, 201-205.

Sierra, I., 1973. Aportación al estudio del cruce Blanco Belga x Landrace: caracteres productivos, calidad de la canal y de la carne. Instituto de Economía y Producciones Ganaderas del Ebro, Volume 16, 43 pp.

Sierra, I. and C. Sañudo, 1987. Qualité de la carcase et de la viande des agneaux et chevreaux de lait dans les pays mediterranens. 38th EAAP Congress, 993.

Talmat, A., G. Monin, M. Briand, M. Dadet and Y. Briand, 1986. Activities of metabolic and contractile enzymes in 18 bovine muscles. Meat Science, Volume 18, 23-40.

Teixeira, A., S. Batista, R. Delfa and V. Cadavez, 2005. Lamb meat quality of two breeds with protected origin designation. Influence of breed, sex and live weight. Meat Science, Volume 71, 530-536.

Velasco, S., S. Lauzurica, V. Cañeque, C. Pérez, F. Huidobro, C. Manzanares and T. Díaz, 2000. Carcass and meat quality of Talaverana breed sucking lambs in relation to gender and slaughter weight. Animal Science, Volume 70, 253-263.

Velasco, S., V. Cañeque, S. Lauzurica, C. Pérez and F. Huidobro, 2004. Effect of different feeds on meat quality and fatty acid composition of lambs fattened at pasture. Meat Science, Volume 66, 457-465.

Vergara, H. and L. Gallego, 1999. Effect of type of suckling and length of lactation period on carcass and meat quality in intensive production systems. Meat Science, Volume 53, 211-215.

Vergara, H., A., Molina and L. Gallego, 1999. Influence of sex and slaughter weight on carcass and meat quality in light lambs produced in intensive systems. Meat Science, Volume 52, 221-226.

Vergara, H. and L. Gallego, 2000. Effect of electrical stunning on meat quality lamb. Meat Science, Volume 56, 345-349.

Wheeler, T.L. and M. Koohmaraie, 1994. Prerigor and postrigor changes in tenderness of ovine Longissimus muscle. Journal of Animal Science, Volume 72, 1232-1238.

Image analysis, ultrasounds and colour reflectance for beef quality estimation

Maria Jose Beriain, K. Insausti and G. Indurain
Escuela Técnica Superior de Ingenieros Agrónomos, Universidad Pública de Navarra, Campus Arrosadía, 31006 Pamplona (Navarra), Spain

Abstract

Classification systems are marketing instruments that try to predict the commercial value of carcasses depending on differences that determine the economic value based on carcass quality categories or commercial cuts. However, meat quality is determined by properties that are perceived by the senses. Developing a method for predicting meat quality characteristics at the slaughterhouse could help the meat industry to ensure beef quality. The present chapter will therefore focus on the application of different technologies in order to measure marbling, tenderness and colour in beef. Ultrasound offers the ability to rapidly and economically estimate carcass characteristics of live animals. Another application has been to study the relationship between fatty acid profile and ultrasound scans. Live animal scanning is said to be better related to carcass and meat traits than carcass scanning because changes occur in muscle during slaughter and ageing. Regarding cattle, image analysis has been used to determine carcass weight, carcass classification and also to estimate meat quality parameters and meat composition relating the data obtained by image analysis with the data determined by chemical analysis. Regarding colour, the regression analysis showed that beef texture at 7 days can be predicted quite accurately just 45 min after slaughter. Regarding colour stability, the best regression equation to predict colour stability in beef aged 7 days included a* measured on the carcass RA muscle 45 min *post-mortem*. It will be necessary to test the suitability of the predictive models developed. But the presented researches revealed that variables easily measured during carcass dressing and beef retail fabrication could complement the current official carcass grading system in a way that carcass quality could accurately reflect meat quality.

Keywords: beef quality estimation, image analysis, ultrasound, colour

Introduction

Classification systems are marketing instruments that try to predict the commercial value of carcasses depending on differences that determine the economic value, such as yield or the quantity of edible meat, or the sensory quality expected by consumers. Most beef classification systems in the world are based on carcass quality categories or commercial cuts. In these systems, visual parameters are assessed because they are easily applied at the slaughter process and these parameters are interesting for livestock producers, retailers and butchers. Meat quality is determined by properties that are perceived by the senses: colour, texture, juiciness and flavour. These properties are not easily assessed by a visual approach of the carcass and they are related to human perception, so that they can change depending of geographic, commercial and psychologist aspects. In this sense, consumers are more and more aware of the difference between intrinsic and extrinsic quality factors and, as a consequence, this fact influences their intention to buy. So, there is a wide range of aspects that can have an effect on the selection of products from consumers when buying (Henson and Northen, 2000; Caswell, 2001).

Falling beef consumption in recent years has led the meat industry to develop 'quality brands' that offer a product with an assured level of quality, thereby enhancing consumer satisfaction.

Developing a method for predicting meat quality characteristics at the slaughterhouse could help the meat industry to ensure beef quality. Different methods have therefore been studied with a view to achieving this goal. In this sense, marbling has been related to texture (Wheeler *et al.*, 1994; Dransfield, 1994), aroma and flavours (Dupuy *et al.,* 1987; Savell and Cross, 1988), and to the effect that meat consumption has on human health due to the relationship between meat fatnesss and fatty acid profile (Eichhorn *et al.*, 1986; Zembayashi *et al.*, 1995). Wulf *et al.* (1997) also tried to determine whether objective measures of muscle colour could be used to classify beef carcasses with regard to tenderness. Besides, colour itself is very important because discolouration of meat cuts can result in a large monetary loss to the retailer, and to other segments of the industry. Hence, it would be very interesting for the meat industry to identify carcass parameters that measured at the early *post-mortem* time can discriminate the shelf life of meat after ageing.

The present chapter will therefore focus on the application of different technologies in order to measure marbling, tenderness and colour. The main characteristics of the instruments that are used to reach this objective are: safety, accuracy, speed and durability (Cross and Whittaker, 1992).

Beef quality estimation

Image analysis

In 1981 Kansas State University had already developed the first-generation video imaging analysis prototype. In Europe, the first trials started in 1980 in Denmark and France. The image analysis system is based on the digitalization of the images that have been taken with a video that has an optic objective adapted to it (photographic, microscopic, etc.). Digitalization converts the saved image in a spot matrix that is identified by a computer depending on position coordinates, lightness and colour (Swatland, 1995).

Regarding cattle, image analysis has been used to determine carcass weight (Basarab *et al.*, 1997; Kuchida *et al.*, 1994), yield grade in the USDA carcass classification system (Belk *et al.*, 1998; Cannell *et al.*, 1999), or conformation in the European carcass classification system, being able to reproduce the behaviour of a grader (Figure 1) (Borggaard *et al.*, 1996; Goyache *et al.*, 2001; Mendizabal *et al.*, 2005). But this technique has also shown high accuracy to estimate meat quality parameters such as composition (Kuchida *et al.*, 2000, Madsen *et al.*, 1996; Standford *et al.*, 1998), colour (Gerrard *et al.*, 1996; Belk *et al.*, 2000), marbling (Figure 2) (Newman, 1987; Gerrard *et al.*, 1996) and meat texture (Shackelford *et al.*, 1998; Li *et al.*, 1999, Wheeler *et al.*, 2003). Nowadays, new video imaging system prototypes have been developed and their introduction on slaughter and packing lines is being evaluated (Cannell *et al.,* 1999, 2002; Belk *et al.*, 2000; Vote *et al.*, 2003; Wyle *et al.*, 2003).

Horgan *et al.* (1995) showed that image analysis can provide accurate data in order to classify carcasses. Mendizabal *et al.* (2005) applied this technology to a wide range of carcasses classified from S to P for conformation (European carcass classification system) in cattle from different productive purposes and breeds. In a second essay, image analysis was applied to carcasses obtained from beef breed purpose yearling bulls. When comparing results from both essays it could be stated that image analysis applied at the slaughter house explained around 90% of conformation variability (using the European grading system) in young bulls. However, the explained variability was 69% in the first essay. These results make in evidence that image analysis offers good perspectives in order to classify carcasses, but also, that these new technologies have to be adapted to each production system.

As it has already been mentioned, image analysis can also be used to estimate meat composition. One of the first studies that applied image analysis to measure meat marbling was carried out by

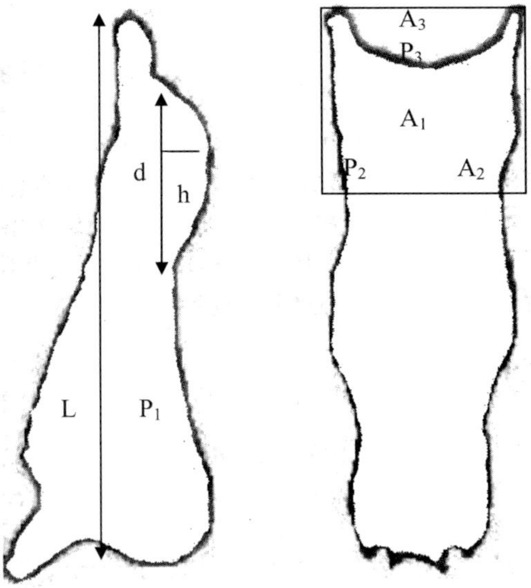

Figure 1. Measurements taken on a bovine carcass (lateral and dorsal) related to conformation (Mendizabal et al.*, 2005).*

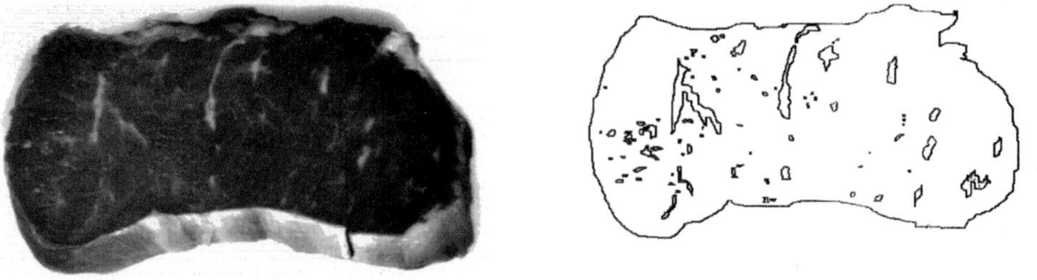

Figure 2. Marbling measured by image analysis (Goñi et al.*, 1999).*

Newman (1984). Due to the different lightness or grey level that can be observed between fat (high lightness) and muscle (low lightness) in digitalized images from the *longissimus dorsi* muscle, marbling can be measured (Figure 2). When relating the data obtained by image analysis with the data determined by chemical analysis a high correlation coefficient (r=0.94) was observed by this same author (Newman, 1987). Besides, Gerrard *et al.* (1996) related these instrumental measurements with those from a visual sensory panel and they reported a close relationship too (r=0.92). However, Ishii *et al.* (1992) found a significant but low relationship between marbling measurements obtained by image analysis and chemical fat content (Soxhlet) (r=0.34). They suggested that fat content would vary in the same sample depending upon location. Therefore, it was thought that this variation in location made the correlation low and measurement using only one location could not accurately estimate the fat content. Besides, marbling is the visual flecks of intramuscular fat, while a chemical solvent sweeps out both visual and non-visual fat from cell frame.

In Japan, beef marbling is considered one of the most important factors that determine beef eating quality and high marbled beef is usually consumed (Zembayashi and Lunt, 1995; Dubeski *et al.*, 1997). In Japan, the ratio marbling area/*longissimus* dorsi area (MA/LDA) was around 8% for the Japanese Black and 5% for the Japanese Shorthorn (Kuchida *et al.*, 1992). At the USA, Gerrard *et al.* (1996) reported 5.28% of intramuscular fat for a sample representative of the US market. These data differ from the ones observed in Spanish local cattle breeds that showed MA/LDA ratios between 0.9% for Asturiana and 2.9% for Avileña breeds in young bulls slaughtered at 300 kg live weight. When slaughtered at 550 kg live weight these values were 2.2% for Asturiana and 6% for Retinta (Goñi *et al.*, 1999). Finally, MA/LDA value was 1.5% in fighting bulls that are traditionally thought to be too fat (Mendizabal *et al.*, 1998). So, the ratio MA/LDA measured by image analysis could estimate the eating quality of beef determined by its intramuscular fat content.

Ultrasounds

Ultrasound has been used for over 40 years to determine body composition in live animals (Stouffer *et al.*, 1959). Ultrasound offers the ability to rapidly and economically estimate carcass characteristics of live animals. Generally, most researches have found ultrasound to estimate carcass back fat thickness with an acceptable degree of accuracy (Brethour, 1992). Results for *longissimus* muscle area have been less conclusive (Smith *et al.*, 1992). Development of transducer designed specifically for cattle use that would allow for imaging for the entire *longissimus* muscle area, has resulted in improved accuracy of this traits (Figure 3) (Herring *et al.*, 1994).

Ultrasound has been in general identified as the technology with the greatest potential for beef because it can be applied to the live animal (Whittaker *et al.*, 1992; Herring *et al.*, 1994; Williams *et al.*, 1997; Wall *et al.*, 2004). However, there is a lack of recent literature comparing accuracy of real time ultrasound measurements taken at ages other than shortly before harvest (Wall *et al.*, 2004).

Another application of ultrasounds has been to study the relationship between fatty acid profile and ultrasound scans. Huerta-Leidenz *et al.* (1993) showed a linear relationship between average of three ultrasound measurements of subcutaneous fat in live animals (at the rib-lumbar area over the 12th-13th ribs, at the rump area and at the perianal region) and monounsaturated/saturated fatty acid ratio (MUFA/SFA ratio).

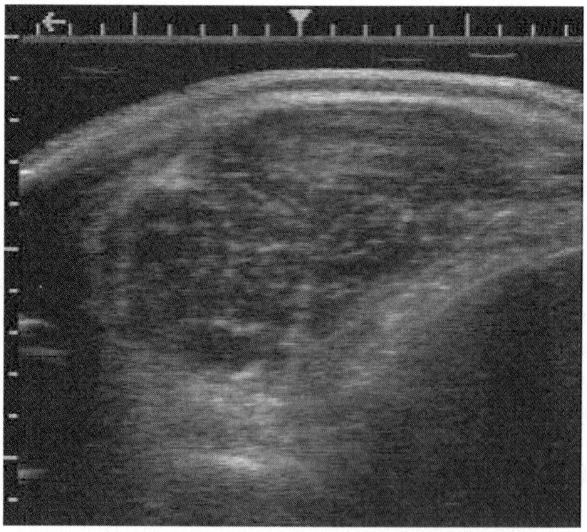

Figure 3. Ultrasound scan where it is shown the longissimus dorsi muscle at the 12th rib.

Indurain *et al.* (2006), scanning beef carcasses, found a significant and positive r-value between marbling and grey level of ultrasonic image scanning at 12th rib (UGL_{12})($r = 0.63$; $p < 0.001$). Besides, there were specific fatty acids that were correlated to ultrasound grey level (Indurain *et al.*, 2006). In intramuscular fat, C18:1 ($r = 3.37$, $p < 0.05$) and MUFA ($r = 0.36$; $p < 0.05$) percentage were positively correlated to whiteness of ultrasonic image scanning at 12th rib (UGL_{12}). However, UGL_{12} had a negative correlation coefficient with C18:2n-6 ($r = -0.51$, $p < 0.01$) and 18:3n-6 ($r = -0.37$, $p < 0.05$). In subcutaneous fat, C18:2n-6 percentage was negatively related to UGL_{12} ($r = -0.34$, $p < 0.05$). C18:0 ($r = 0.34$, $p < 0.05$) and SFA ($r = 0.34$; $p < 0.05$) percentage were positively related to the ultrasonic measured area of the *longissimus* muscle at the 12th rib ($ULMT_{12}$). On the contrary, C16:1 ($r = -0.37$, $p < 0.05$), C18:1 ($r = -0.33$, $p < 0.05$) and MUFA ($r = -0.39$; $p < 0.05$) were negatively correlated to $ULMT_{12}$.

Whittaker *et al.* (1992) suggested the idea that live animal scanning is better related to carcass and meat traits than carcass scanning because changes occur in muscle during slaughter and ageing. This suggests that when this technology will be applied to live animals, better correlation coefficients between ultrasonic readings and fatty acid profile could be expected. In this sense, in one of our recent researches, it was observed that when scanning live animals the best R^2 values were obtained 25 days before slaughter (data not shown). At this moment, the European and the USDA grading systems can be estimated (44 to 69% variability). Besides, 37 to 40% of odour, characteristic beef flavour and acceptability can also be estimated (unpublished data).

Colour reflectance

The *Commission Internationale de l'Eclairage* (CIE) has defined different representation systems in order to specify colour in an objective way. In these systems colour is determined by three coordinates but the recommended one is the CIE L*a*b (or reflectance) system because it shows the higher correlation with visual assessments (Hunt *et al.*, 1991). Spectrophotocolorimeters and colorimeters are the instruments to measure colour that can receive the wave length coming from objects.

Most research has mainly been focused on the use of colour reflectance measured on the carcass to estimate tenderness and colour stability after ageing.

Regarding the estimation of tenderness, Jeremiah *et al.* (1991), Wulf *et al.* (1997), and Tatum *et al.* (1999) reported that objective measurements of *longissimus dorsi* muscle (LD) colour were related closely enough to beef tenderness to allow carcasses to be sorted into carcass groupings of different palatability. Wulf and Page (2000) also found that when introducing L* and b* in the multiple regression equations to predict *longissimus dorsi* shear force and palatability the prediction R^2 was higher than considering only marbling. The *longissimus dorsi* muscle colour would be therefore a non-invasive method and a potentially useful predictor of beef tenderness.

Goñi *et al.* (2007) found that the best colour variable to predict *longissimus dorsi* Warner Bratzler shear force in beef aged seven days (WBSF 7 d) was a cubic equation with b* taken on the sliced *longissimus dorsi* after blooming 60 minutes ($R^2 = 0.19$). However, the measurement of muscle b* presents problems when applied to the industry chain due to its sensitivity to bloom time (Wulf and Wise, 1999). b* value is increasing at a rate of 0.1 to 0.2 unit per min. Therefore, in order to use b* as a predictor of palatability or texture characteristics, bloom would need to be monitored to the exact minute or a minimum bloom time of approximately 78 min would be required to achieve an stable b* value (Wulf and Wise, 1999). However, any stoppage of the bloom chain would require some excellent coordination or a sophisticated computer system to ensure and accurate measure system. Because of the measured problems associated with muscle b*, a small sacrifice in accuracy of tenderness predictions was made and the best equation in which b* was not included was chosen.

A cubic regression equation in which H* was taken on the *latissimus dorsi* muscle 24 hours post-mortem (H*LTD24h) showed the highest R^2 value (0.17; $p < 0.05$) (Goñi *et al.*, 2007).

Bloom time would still need to be monitored when measuring H*; however, the effect of bloom time on H* (and L* coordinate too) is minimal compare with the effect of bloom time on b* (Wulf and Wise, 1999).

The best regression equation to predict sensorial texture of beef aged seven days (TEX 7d) was a cubic equation which includes C* coordinate of *rectus abdominus* measured 45 minutes after slaughter (C*RA45min)(R^2=0.17; $p < 0.05$), being TEX a texture index = initial juiciness + sustained juiciness + easiness to swallow + lubricity – hardness – cohesiveness – WBSF) (Goñi *et al.*, 2007).

Similarly to b* coordinate, C* (and a* values too) depends on blooming time and would achieve a stable value after 24h hours of air exposure when beef oxygenation is completed and meat has achieved its final pH value (MacDougall, 1994).

With regard to beef aged 14 days the best regression equations included a* and C* coordinates measured on meat after a short time of exposure to air. After 24h hours of air exposure, the best regression equations for both tenderness parameters were cubic regression equations which included the b* coordinate of *lattisimus dorsi* muscle measured 24 hours post-mortem (Goñi *et al.*, 2007).

The derived regression equations explained approximately between 20 and 40 per cent of WBSF and TEX variability. To date, instrumental readings of the colour of certain carcass muscles have not proved to be an entirely satisfactory means of classifying carcasses according to the WBSF and TEX values of the meat as illustrated in this study and in other similar studies (Wulf *et al.,* 1997). A series of other factors such as fatness (Li *et al.,* 1999; Wulf and Page, 2000; Wheeler *et al.*, 2002; Vote *et al.*, 2003) or pH (Wulf and Page, 2000) play a role and are additive in accounting for variations in beef texture. However, the narrow range of carcass fatness and pH in the carcasses used in the study by Goñi *et al.* (2007) failed to improve the predictive value of these variables together with colour coordinates.

In summary, carcass surface muscle (LTD and RA) colour on the day of slaughter (45min) would have a low relationship with WBSF (7d and 14d) and the TEX (7d and 14d) in beef. On the contrary, the colour measurements taken 24 h *postmortem* on the LTD, RA and LD muscles were correlated with WBSF and with the TEX (Goñi *et al.*, 2007). So, it seems clear that more information on beef texture is obtained with colour measurements at 24 h. However, the regression analysis showed that beef texture at 7 days can be predicted quite accurately just 45 min after slaughter. This fact could very important for the meat industry in Spain because most beef quality labels recommend that beef meat should be aged 7 days.

Regarding colour stability, it is regulated by many factors that are inter-related (oxygen availability, enzymatic systems, mitochondrial activity, muscle integrity, lipid oxidation, microbial spoilage, etc.). In beef from young bulls, these factors provide meat with an acceptable colour for consumers after 1 h blooming. On the contrary, after 48 h display a high percentage of beef samples might be rejected by consumers due to discolouration problems and leading to an important economic loss (Insausti *et al.*, 2006). Due to the relationship between metmyoglobin (MMb) and the acceptability of meat by consumers, the MMb percentage was used to segregate between acceptable and rejectable beef colour in the present study. MMb percentage was also used to estimate which are the samples that are more likely bound to undergo discolouration, which is of great interest for the meat industry because it is necessary that carcass classification can be able to predict at the slaughter house the sensory quality of a product that will be consumed some days after slaughter. In this sense, the best regression equation to predict colour stability (expressed as MMb % after 48h display) in beef aged 7 days included a* measured on the carcass RA muscle 45 min *post-mortem* ($R2 = 0.38$, $p < 0.001$, unpublished data).

Conclusions

Being able to predict meat texture, colour stability and fatty acid composition so that beef can be offered to consumers in optimal conditions, while avoiding unnecessary refrigeration costs, would of course be of interest to the meat industry. At the same time, bringing a more consistent product to market should increase consumer satisfaction with beef.

It will be necessary to test the suitability of the predictive models developed and to improve them by increasing the number of observations and by combining them with other supplementary methods, as well as to apply them to other animals of different breeds, ages, and sex, and thus degree of marbling.

The presented researches revealed that variables easily measured during carcass dressing and beef retail fabrication, such as carcass weight, fat thickness, or marbling estimation by image analysis, ultrasound readings or colour reflectance could complement the current official carcass grading system and achieve a more effective carcass sorting procedure in very local and specific productions systems, in which carcass quality could accurately reflect meat quality.

References

Basarab, J.A., D. Milligan, J.J. McKinnon and B.E. Thorlakson, 1997. Potential use of video imaging and real-time ultrasound on incoming feeder steers to improve carcass uniformity. Canadian Journal of Animal Science, Volume 77, 385-392.

Belk, K.E., J.A. Scanga, J.D. Tatum, J.W. Wise and C.G. Smith, 1998. Simulated instrument augmentation of USDA yield grade application to beef carcasses. Journal of Animal Science, Volume 76, 522-527.

Belk, K.E., J.A. Scanga, A.M. Wyle, D.M. Wulf, J.D. Tatum, J.O. Reagan and G.C. Smith, 2000. The use of video image analysis and instrumentation to predict beef palatability. Proceeding 53rd Annual Reciprocal Meat Conference of the American Meat Science Association, Savory, IL, USA, 10-15.

Borggaard, C., N.T. Madsen and H.H. Thodberg, 1996. In-line image analysis in the slaughter industry, illustrated by beef carcass classification. Meat Science, Volume 43, S151-163.

Brethour, J.R. 1992. The repeatability and accuracy of ultrasound in measuring back fat thickness. Journal of Animal Science, Volume 70, 1039-1044.

Cannell, R.C., K.E. Belk, J.D. Tatum, J.W. Wise, P.L. Chapman, J.A. Scanga and G.C. Smith, 2002. Online evaluation of a commercial video image analysis system (Computer Vision System) to predict beef carcass red meat yield and for augmenting the assignment of USDA yield grades. Journal of Animal Science, Volume 80, 1195-1201.

Cannell, R.C., J.D. Tatum, K.E. Belk, J.W. Wise, R.P. Clayton and G.C. Smith, 1999. Dual-component video image analysis system (VIASCAN) as a predictor of beef carcass red meat yield percentage and for augmenting application of USDA yield grades. Journal of Animal Science, Volume 77, 2942-2950.

Caswell J.A., 2001. Food Safety and the U.S. Consumer. Proceedings 71st EAAE Seminar The Food Consumer in the Early 21st Century. Zaragoza, Spain.

Cross, H.R. and A.D. Whittaker, 1992. The role of instrument grading in a beef value-based marketing system. Journal of Animal Science, Volume 70, 984-989.

Dransfield, E. 1994. Tenderness of meat, poultry and fish. In: Pearson, A. M. and T.R. Dutson (eds.), Quality attributes and their measurement in meat, poultry and fish products. Blackie Academic and Professional, London, UK, 289-315.

Dubeski P.L., D.M. Jones, J.L. Aalhus and W.M. Robertson, 1997. Canadian, American, and Japanese carcass grades of heifers fed to heavy weights to enhance marbling. Canadian Journal of Animal Science, Volume 77, 393-402.

Dupuy, H.P., M.E. Bailey, A.J. St Angelo, J.R. Vercellotti and M.G. Legendre, 1987. Instrumental analysis of volatiles related to warmed-over flavour of cooked meats. In: St Angelo, A. J. and M.E. Bailey (eds.), Warmed-over flavour of meat. CRC Academic Press Inc., Boca Raton, FL, USA, 165-191.

Eichhorn, J.M., L.J. Coleman, E.J. Wakayama, G.J. Blomquist, C.M. Bailey and T.G. Jenkins, 1986. Effects of breed type and restricted versus ad libitum feeding on fatty acid composition and cholesterol content of muscle and adipose tissue from mature bovine females. Journal of Animal Science, Volume 63, 781-794.

Gerrard, D.E., X. Gao and J. Tan, 1996. Beef marbling and color score determination by image analysis. Journal of Food Science, Volume 61, 145-147.

Goñi M.V., M.J. Beriain, G. Indurain and K. Insausti, 2007. Predicting longissimus dorsi characteristics in beef based on early post-mortem colour measurements. Meat Science, Volume 76, 38-45.

Goñi, V., J.A. Mendizabal, J.M. Beriain, P. Alberti, A. Arana, A.P. Aguinoa and A. Purroy, 1999. Marbrure de la viande de veaux de sept races á viande espagnoles determinée par analyse d'image. Proceedings of the 6th Rencontres Recherches Ruminants, Paris, France, pp 278.

Goyache, F., A. Bahamonde, J. Alonso, S. Lopez, J. Alonso, J.J. del Coz, J.R. Quevedo, J. Ranilla, O. Luaces, I. Alvarez, L. Royo and J. Diez, 2001. The usefulness of Artificial Intelligence techniques to assess subjective quality of products in the food industry. Trend in Food Science and Technology, Volume 12, 370-381.

Henson, S. and J. Northen, 2000. Consumer assessment of the Safety of Beef at the Point of Purchase: a Pan-European Study. Journal of Agricultural Economics, Volume 51, 90-105.

Herring, W.O., D.C. Miller, J.K. Bertrand and L.L. Benyshek, 1994. Evaluation of machine, technician and interpreter effects on ultrasonic measures of backfat and longissimus muscle area in beef cattle. Journal of Animal Science, Volume 72, 2216-2226.

Horgan G.W., S.V. Murphy and G. Simm, 1995. Automatic assessment of sheep carcasses by image analysis. Animal Science, Volume 60, 197-202.

Huerta-Leidenz, N.O., H.R. Cross, J.W. Savell, D.K. Lunt, J.F. Baker, L.S. Pelton and S.B. Smith, 1993. Comparison of the fatty acid composition of subcutaneous adipose tissue from mature Brahman and Hereford cow. Journal of Animal Science, Volume 71, 625-630.

Hunt, M.C., J.C. Acton, R.C. Benedict, C.R. Calkins, D.P. Cornforth, L.E. Jeremiah, D.G. Olson, C.P. Salen, J.W. Savell and S.D. Shivas, 1991. Guidelines for meat color evaluation. Proceedings 44th Reciprocal Meat Conference of the American Meat Science Association, Chicago, IL, USA, 3-14.

Indurain G., M.J. Beriain, M.V. Goñi, A. Arana and A. Purroy, 2006. Composition and estimation of intramuscular and subcutaneous fatty acid composition in Spanish young bulls. Meat Science, Volume 73, 326-334.

Insausti, K., G. Indurain, M.V. Goñi, V. Sarriés, B. Hernández and M.J. Beriain, 2006. Colour stability of beef stored under vacuum packaging and its relationship to carcass fatness. Proceedings of 52nd ICoMST, Dublin, Ireland, 525-526.

Ishii T., R.G. Cassens, K.K. Scheller, S.C. Arp and D.M. Schaefer, 1992. Image analysis to determine intramuscular fat in muscle. Food Structure, Volume 11, 55-60.

Jeremiah, L.E., A.K.W. Tong and L.L. Gibson, 1991. The usefulness of muscle color and pH for segregating beef carcasses into tenderness groups. Meat Science, Volume 30, 97-114.

Kuchida K., K. Yamaki, T. Yamagishi and Y. Mizuma, 1992. Evaluation of meat quality in Japanese beef cattle by computer image analysis. Animal Science and Technology (Jpn.), Volume 63, 121-128.

Kuchida, K., S. Kono, K. Konishi, L.D. Van Vleck, M. Suzuki and S. Miyoshi, 2000. Prediction of crude fat content of longissimus muscle of beef using the ratio of fat area calculated from computer image analysis: comparison of regression equations for prediction using different input devices at different stations. Journal of Animal Science, Volume 78, 799-803.

Kuchida, K., T. Yamagishi, H. Takeda and K. Yamaki, 1994. The estimation of beef carcass traits using computer image analysis for body measurement in Japanese Black steers. Animal Science and Technology, Volume 65, 401-406.

Li, J., J. Tan, F.A. Martz and H. Heymann, 1999. Image texture features as indicators of beef tenderness. Meat Science, Volume 53, 17-22.

MacDougall, D.B. 1994. Colour of meat. In: Pearson, A.M. and T.R. Dutson (eds.), Quality Attributes and their measurement in meat, poultry and fish products. Advances in meat research. Blackie Academic & Professional, Glasgow, UK, pp 79.

Madsen, N.T., H.H. Thodberg, T. Flig and E. Ovesen, 1996. BCC-2 for objective beef carcass classification and prediction of carcass composition. Proceedings 42nd ICoMST, Lillehammer, Norway, 244-245.

Mendizabal J.A., A. Purroy, M.J. Beriain, G. Lizaso and K. Insausti, 1998. Medida del grado de veteado de la carne mediante Análisis de Imagen. Caso de la carne de toro de Lidia, Información Técnica Económica Agraria, Volume 94ª (I), 43-48.

Mendizabal, J.A., A. Legarra and A. Purroy, 2005. Relación entre la nota de conformación y diferentes medidas morfométricas realizadas mediante análisis de imagen en canales bovinas de diferentes conformaciónes. Información Técnica Económica Agraria, Volume 26 (II), 765-767.

Newman P.B., 1984. The use of video image analysis for quantitative measurement of visible fat and lean in meat: part I-Boneless fresh and cured meats. Meat Science, Volume 10, 87-100.

Newman, P.B., 1987. The use of video image analysis for quantitative measurement of visible fat and lean in meat: Part III-Lipid content variation in commercial processing beef and its prediction by Image Analysis. Meat Science, Volume 19, 129-137.

Savell, J.W. and H.R. Cross, 1988. The role of fat in the palatability of beef, pork and lamb. In: Designing foods: animal product options in the marketplace, National Academic Press, Washington DC, USA, 345-355.

Shackelford, S.D., T.L. Wheeler and M. Koohmaraie, 1998. Coupling of image analysis and tenderness classification to simultaneously evaluate carcass cutability, longissimus area, subprimal cut weights, and tenderness of beef. Journal of Animal Science, Volume 76, 2631-2640.

Smith, M.T., J.W. Oltjen, H.G. Dolezal, D.R. Gill, and B.D. Behrens, 1992. Evaluation of ultrasound for prediction of carcass fat thickness and longissimus muscle area in feedlot steers. Journal of Animal Science, Volume 70, 29-37.

Stanford, K., R.J. Richmond, S.D.M. Jones, W.M. Robertson, M.A. Price and J. Gordon, 1998. Video image analysis for on-line classification of lamb carcases. Animal Science, Volume 67, 311-316.

Stouffer, J.R., M.V. Wellentine and G.H. Wellington, 1959. Ultrasonic measurement of fat thickness and loin eye area on live cattle and hogs. Journal of Animal Science, Volume 18, 1483.

Swatland, H.J., 1995. Video image analysis, In: On-Line evaluation of meat, Technomic Publishing Company Inc., Lancaster, PA, USA, 271-290.

Tatum, J.D., K.E. Belk, M.H. George and G.C. Smith, 1999. Identification of quality management practices to reduce the incidence of retail beef tenderness problems: development and evaluation of a prototype quality system to produce Tender beef. Journal of Animal Science, Volume 77, 2112-2118.

Vote, D.J., K.E. Belk, J.D. Tatum, J.A. Scanga and G.C. Smith, 2003. Online prediction of beef tenderness using a computer vision system equipped with a BeefCam module. Journal of Animal Science, Volume 81, 457-465.

Wall, P.B., G.H. Rouse, D.E. Wilson, R.G. Tait Jr., and W.D. Busby, 2004. Use of ultrasound to predict body composition changes in steers at 100 and 65 days before slaughter. Journal of Animal Science, Volume 82, 1621-1623.

Wheeler, T.L., L.V. Cundiff and R.M. Koch, 1994. Effect of marbling degree on beef palatability in Bos Taurus and Bos Indicus cattle. Journal of Animal Science, Volume 72, 3145-3151.

Wheeler, T.L., D. Vote, J.M. Leheska, S.D. Shackelford, K.E. Belk, D.M. Wulf, B.L. Gwartney and M. Koohmaraie, 2002. The efficacy of three objective systems for identifying beef cuts that can be guaranteed tender. Journal of Animal Science, Volume 80, 3315-3327.

Wheeler, T.L., D. Vote and J.M. Leheska, 2003. The efficacy of three objective systems for identifying beef cuts that can be guaranteed tender. Journal of Animal Science, Volume 80, 3315-3327.

Whittaker, A.D., B. Park, B.R. Thane, R.K. Miller and J.W. Savell, 1992. Principles of ultrasound and measurement of intramuscular fat. Journal of Animal Science, Volume 70, 942-952.

Williams, R.E., J.K. Bertrand, S.E. Williams and L.L. Benyshek, 1997. Biceps femoris and rump fat as additional ultrasound measurements for predicting retail product and trimmable fat in beef carcasses. Journal of Animal Science, Volume 75, 7-13.

Wulf, D.M. and J.K. Page, 2000. Using measurements of muscle color, pH, and electrical impedance to augment the current USDA beef quality grading standards and improve the accuracy and precision of sorting carcasses into palatability groups. Journal of Animal Science, Volume 78, 2595-2607.

Wulf, D.M. and J.W. Wise, 1999. Measuring muscle color on beef carcasses using the L* a* b* color space. Journal of Animal Science, Volume 77, 2418-2427.

Wulf, D.M., S.F. O'Connor, J.D. Tatum and G.C. Smith, 1997. Using objective measures of color to predict beef longissimus tenderness. Journal of Animal Science, Volume 75, 685-692.

Wyle, A.M., D.J. Vote, D.L. Roeber, R.C. Cannell, K.E. Belk, J.A. Scanga, M. Goldberg, J.D. Tatum and G.C. Smith, 2003. Effectiveness of the SmartMV prototype BeefCam System to sort beef carcasses into expected palatability groups. Journal of Animal Science, Volume 81, 441-448.

Zembayashi, M. and D.K. Lunt, 1995. Distribution of intramuscular lipid throughout M. longissimus thoracis and lumborum in Japanese Black, Japanese Shorthorn, Holstein and Japanese Black crossbreds. Meat Science, Volume 40, 211-216.

Zembayashi, M., K. Nishimura, D.K. Lunt and S.B. Smith, 1995. Effect of breed type and sex on the fatty acid composition of subcutaneous and intramuscular lipids of finishing steers and heifers. Journal of Animal Science, Volume 73, 3325-3332.

Spectral and thermal imaging for meat quality evaluation

Paolo Menesatti, S. D'Andrea and C. Costa
CRA-ISMA – Istituto Sperimentale per la Meccanizzazione Agrícola del Consiglio per la Ricerca in Agricoltura, Via della Pascolare 16, 00016 Monterotondo (Roma), Italy

Abstract

The analysis of meat optical characters is assuming to have a great relevance in product and organoleptic quality assessment. Spectrophotometric techniques are associated to a high analytic ability (multiparametric and multichannel), interesting properties of non destructiveness measurements, relative simplicity, rapidity, portability in operative environments, informatics and electronics interface with other instrumental sensors. Digital imaging based on spectrophotometry and infrared thermal imaging (equipment, architecture set-up and measurements) will be explained. Hyperspectral imaging procedure and statistical approaches will be deepen. Three practical examples were conducted: the first on spectral visual inspection on meat slices with punctual and multi spectral approaches, the second on hyperspectral image analysis on beef meat and the third on infrared thermal imaging on sheep carcass.

Keywords: meat quality, photonics, optics, electronics, PLS

Introduction

The analysis of meat optical characters is assuming to have a great relevance in product and organoleptic quality assessment. Spectrophotometric applications are relevant, above all, in particular because outputs from meat under light beam for single spectral band and components, could contribute to a more detailed and refined evaluation of quality characteristics, indicating, where possible, important information on production and origin. Spectrophotometric techniques are associated to a high analytic ability (multiparametric and multichannel), interesting properties of non destructiveness measurements, relative simplicity, rapidity, portability in operative environments, informatics and electronics interface with other instrumental sensors. For this reason, in the last few years, the scientific and applicative interests considerably increase toward spectrophotometric applications on quality aspects of meat.

Advanced technologies (imaging spectrophotometers) are able to acquire spectral images with a good optical resolution (150-250 kpixel) where each pixel own the whole spectral information (visible and near infra-red). Multi- (≤ 10 spectral bands) or hyper- (> 10 spectral bands) spectral imaging analysis is the new frontier of optical imaging on meat with a high capability of a more detailed analysis on product structure and quality. In the next few years the technological evolutions of photonics will reach a break-even point where this system could be adopted, for their low price, on line for better product qualification and their safety of use (Chen, 1992; Park *et al.*, 2004; Yang *et al.*, 2005).

Some authors applied near-infrared reflectance for the determination of fat, protein and moisture in fresh pork or beef meat (Kruggel *et al.*, 1981; Martens *et al.*, 1981; Lanza, 1983; Josell *et al.*, 2000; Swatland, 1997; Geesink, 2003; Irie and Iwaki, 2003; Leroy *et al.*, 2003; Menesatti *et al.*, 2003). Infrared spectroscopy allowed also multi-component analysis of meat products (Bjarno *et al.* 1982) and the prediction of beef tenderness (Jeyamkondan *et al.*, 2004).

Some interesting spectral application were found for the poultry industry as: carcasses analysis (Park and Chen, 2001), detection of surface contaminants (Lawrence *et al.*, 2004), separating septicemic and normal chicken livers (Hsieh *et al.*, 2002), assessment in chicken breast (Tao and Ibarra, 2001), chicken heart disease characterization (Chao *et al.*, 2001), detection of chicken skin tumours (Chao *et al.*, 2002) and classification diseased (Chen, 1992). Some other application concerns multi-spectral analysis (Skjervold *et al.*, 2003; Yang *et al.*, 2006).

The infrared thermal imaging is another interesting 'non conventional' imaging technique. Infrared thermometry is a non-invasive, non-destructive method that is based on measuring specific electromagnetic radiation emitted by any object, according to the Stefan-Boltzman's and Planck's laws (see Maldague, 1994; Rahkonen and Jokela, 2003 for basic theory).

All objects emit heat (energy) waves. If an object is cold, its molecules vibrate slower and energy of longer wavelengths is emitted. When the temperature of the object rises, its molecules vibrate faster and the wavelength becomes shorter. Every particular energy wavelength has a temperature associated with it. The lowest temperature that the human eye can visually detect is approx. 500 °C. At temperatures cooler than that, the heat waves emitted are invisible and are called *infrared* waves. Infrared light comprises the lower frequency energy emissions which are undetectable by the human eye. Our eyes are not sensitive to such low-energy radiation, but IR can reveal useful information about our environment. An infrared camera can see the 'invisible' light and convert it to a video signal, which can be seen on a television or on the eye-piece of the camera, and can display the temperature associated with of any part of that image.

This technology, linked with a visual image analysis or automated, could better evaluate thermal phenomenon mostly on carcass, differentiating the analysis for areas and zones. For the readiness of the thermal camera use, thermal image analysis on carcass could represent a valid inspection system and a control method for the characteristics and the process of storage post-slaughtering. Infrared thermal imaging was applied to evaluate cooked and raw chicken meat (Ibarra *et al.*, 2000) and to study the animal welfare (Menesatti *et al.*, 2004).

The following paragraphs will deepen both technological / methodological and analytical aspects of spectral / thermal imaging for meat quality evaluation with three practical examples.

Digital imaging based on spectrophotometry: equipment, architecture set-up and measurements

Imaging spectrophotometry is based on the utilization of an integrated hardware and software architecture (HW&SW) able to digitally capture and handle spectra as an image sequence, as they results along a pre-defined alignment on a surface sample properly energized. According to the different wavelength of the source and to the different device spectral sensitivity, different physical-chemical superficial characteristics of the sample can thus be investigated and analyzed.

The imaging based spectrophotometry architecture set-up, adopted in the following examples, are based on the utilization of the ImSpector™ spectrometer devices family, from SpecIm™ Oy (Figure 1). The basic architecture of the ImSpector™ spectrophotometer devices is constituted by optics based on volume type holographic transmission grating. The grating is used in patented prism-grating-prism construction (PGP element) which can be shown to have high diffraction efficiency, good spectral linearity and it is nearly free of geometrical aberrations due to the on-axis operation principle. The incoming line image (frame) is projected and dispersed onto the CCD, for VIS analysis, or InGaAs array, for NIR analysis. Each frame contains the line pixels in one dimension (spatial axis) and the spectral pixels in the other dimension (spectral axis), providing full spectral information for each

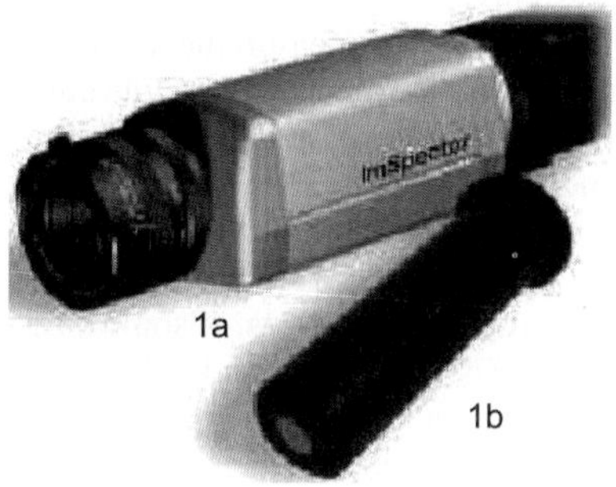

Figure 1. ImSpector™, spectrophotometric device: (a) commercial general purpose configuration, (b) OEM configuration.

line pixel. The reconstruction a whole hyperspectral image of the sample was performed scanning the sample line-by-line as all the field of view defined by the optical system is acquired.

Two different spectrophotometers have been utilized: an ImSpector™ V10, for the analysis performed in the range from 400 nm to 1000 nm (VIS) and an ImSpector™ N17 for the analysis performed in the range from 1000 nm to 1700 nm (NIR). For the ImSpector™ V10, the resolution of the image line was 700 pixels by 10 bits. The number of frames (image resolution in Y axes) can vary from 10 to 500 depending from the speed and the accuracy of the scanner. For the ImSpector™ N17, the resolution of the image line was 320 pixels by 12 bits. The number of frames (image resolution in Y axes) can vary from 10 to 300 depending from the speed and the accuracy of scanning.

The detection architecture system (Figure 2) is basically constituted by four elements: (1) a sample transportation plate (DV™ spectral scanner), (2) a collimated illumination device (Fiber-lite) composed by one 150W halogen lamp, as source light, and one illumination opening in optical fibre of 200 mm long and 2 mm width, positioned at 45° in respect to the transportation plate (i.e. to the sample) and presenting a minimum light divergence, iii) an imaging spectrophotometer, that is an ImSpector™ V10, for the analysis performed in the VIS and an ImSpector™ N17 for the ones performed in the NIR) coupled with a standard C-mount zoom lens and iv) an array detector, that is Teli CCD monochrome camera for the ImSpector™ V10 and Pixelvision InGaAs IR camera for the ImSpector™ N17.

Figure 2. Heart of the NIR imaging spectrophotometer detection system. The spectral scanner, not visible in the picture, allows the synchronized translation of the sample.

Differently from other detection systems, always based on spectroscopy, the detection architecture adopted offers several advantages: (1) reduced measurement time, (2) no scanning movement of the detection units and (3) simultaneous measurement over a line area. In many applications several single points can be replaced by one multi-channel imaging spectrometer thereby saving both in the instrumentation cost and in the amount of mechanics and required space. There are 'disadvantages' too, namely the huge amount of data that should be collected and processed at once and stored for later processing. Therefore, in some process application this may require sophisticated data processors, like DSP (Digital Signal Processors) circuits, or some kind of preliminary data reduction scheme before actual data processing. The latter one can be done electronically or mathematically using fast processors.

All spectral values were expressed in terms of relative reflectance (R), applying the following equation:

$$R = \frac{r_s - r_b}{r_w - r_b} \qquad (1)$$

being R the relative reflectance of the sample at each wavelength, r_s the absolute signal value (radiance) measured for the sample at each wavelength, r_b the absolute signal value (radiance) measured at each wavelength for the black (background noise) and r_w the absolute signal value (radiance) measured at each wavelength for the standard white (100% of reflectance).

Spectral images processing

Spectral images, obtained with the instruments above mentioned, could be processed with two different approaches:

1. spectral visual inspection;
2. multi/hyper-spectral imaging.

Spectral visual inspections allow to analyze specific areas within the object image. Those analyses could be either qualitative or quantitative depending on a need of a clustering classification or a prediction, for example of a substance in a sampling. In both cases it is possible to use spectral images obtained with instruments above descript (NIR and VIS-NIR Scanners); on those images Regions of Interest (ROI) will be selected to obtain spectral information. Obtained values represent mean spectral reflectance between selected pixels.

The Multi- ($\leq$ 10 spectral bands) or hyper- (> 10 spectral bands) spectral imaging is a complex technique that involves automated processing of Visible or Near Infrared spectroscopy images (but also X-ray, Electron energy loss, Raman, cathodoluminescence spectroscopy).

The primary advantages of hyperspectral mapping are that, because an entire spectrum is acquired at each point, the operator needs a no a priori knowledge of the sample and post-processing allows all available information from the dataset to be mined. The primary disadvantage is that fast computers, sensitive detectors, and large data storage capacities are needed. In both cases, due to the large amount of information, characteristic of multichannel instruments, multivariate analysis techniques are utilized to discriminate among classes or predict chemical compounds.

Two multivariate methods of spectra handling are briefly described. The first method, principal components analysis (PCA) is a technique used to summarize data by forming new variables (components), which are linear composites of the original variables. This method could be called 'unsupervised'. The second, Partial Least Squared (PLS) is a soft-modelling method to built predictive models when the factors are many and highly collinear. This method could be called 'supervised' because information on classes are introduced *a priori* to extract the best fitting model.

Principal components analysis (PCA) is a procedure used to find hypothetical variables (components) which explain more of the variance in the multidimensional data as possible (Davis, 1986; Harper, 1999). These new variables are linear combinations of the original ones. The advantages of this method is the reduced dimensionality of input data, removal of ground noise from original data and a reduced possibility of over-fitting the training data. PCA has several applications, two of them are:

- simple reduction of the data set to only two variables (the two most important components), for plotting and clustering purposes;
- the most important components are correlated with some other underlying variables. For morphometric data, this might be simply age, while for associations it might be a physical or chemical gradient (e.g. latitude or position across the shelf).

Partial Least Squares (PLS) represents one of the more reliable chemometric techniques used to look forward correlations among the exit signal of a multi-channel instrument and the information included in the measure itself.

The model works through a specific algorithm (SIMPLS; De Jong, 1993) on the whole array variables (input vars., X matrix) and on the reference values (observed values, Y matrix). Through an initial calculation procedure of maximum correlations, the model determines the minimum set of the n estimation variables (latent and score variable) by a recursive process.

These variables could be represented in a n-dimensional space and used by PLS to calculate the best regression matrix (B) between the X and the Y. The optimum B matrix is the one that minimize the estimation errors matrix of the model (E), according the matrix equation:

$$Y = BX + E \quad (2)$$

The model includes a training phase and a validation phase for both of which it calculates the residual errors (PRESS - Predicted Residual Error Sum of Squares).

Generally, the trend of the residual errors is a decrease in the calibration phase (RMSEC) and increase for the validation phase (RMSECV). The number of the latent variables used by the models is based on the concurrent minimization of these two types of errors.

Through the B matrix, the new values of Y (esteemed values, Y' matrix) are esteemed. The comparison between the observed values (Y) and the esteemed one (Y') allows to evaluate the degree of correct prediction (SEP, standard error of prediction) and to test the reliability of the model, in the mean time.

In many cases, spectral image processing is anticipate by a pre-treatment methods, i.e. normalization, centering, auto-scaling, pareto scaling, range scaling, log transformation and power transformation (Van den Berg *et al.*, 2006).

Hyperspectral imaging

In hyperspectral imaging, images are processed through a complex procedure, briefly described as follows:

1. the 3D matrix (X, Y = spatial information, λ = spectral information) is reduced in one o few distinct 2D matrices (X,Y grey-scaled images) usually applying multivariate techniques;
2. the grey-scaled (8 bit) images are filtered and segmented in B/W (2 bit) images;
3. morphological operators and blob or texture analysis is performed to measure different objects features (count, area, perimeter, shape, etc.).

In many cases, a visual analysis is performed directly on the grey-scaled images to look for possible characteristics or specific textures, not in evidence observing single wavelength images.

To perform the first step in hyperspectral imaging, PCA is normally used. To explain the procedure, we must consider the representation of an image. A square, X by Y image can be expressed as an N^2-dimensional vector:

$$X = (\quad X_1 \quad X_2 \quad X_3 \quad . \quad . \quad X_N) \tag{3}$$

where the rows of pixels in the image are placed one after the other to form a one-dimensional image. E.g., the first X elements (X_1-X_N) will be the first row of the image, the next elements are the next row, and so on. The values in the vector are the intensity values of the image for each λ (wavelength) layer (Figure 3).

If we have 20 λ layers, each layer is X pixels high by Y pixels wide. For each layer we can create an image vector as described in Equation 1. Then we can put all the layers together in one big image-matrix like this:

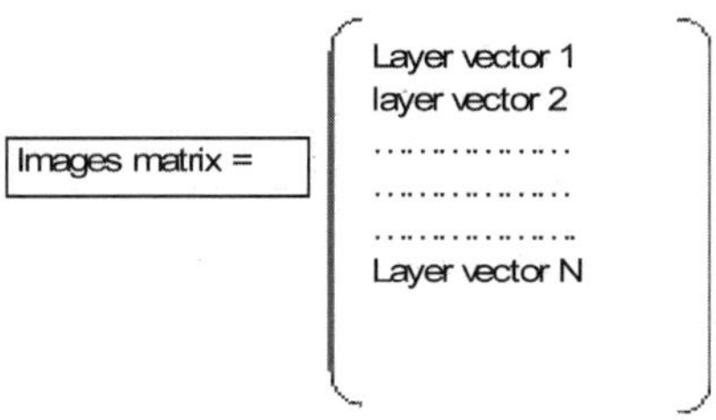

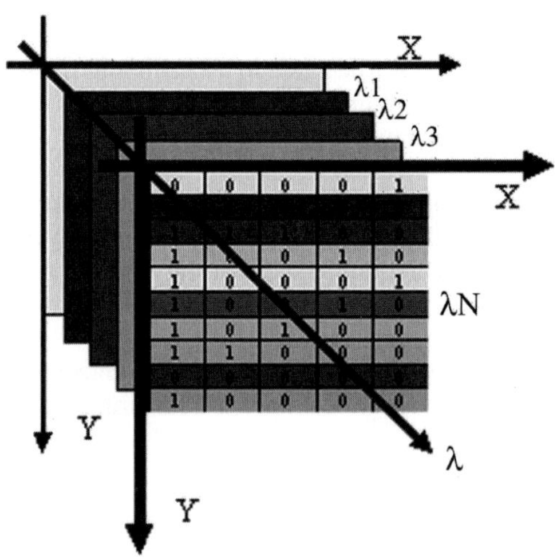

Figure 3. Graphic example of a hyperspectral image.

which gives us a starting point for our PCA analysis. Once we have performed the PCA, we own our original data in terms of the eigenvectors we found from the covariance matrix. The way this is done is computer vision is to measure the difference between both layers and other multiple-layer images (hyperspectral), not along the original axes, but along the new axes derived from the PCA analysis. The PCA analysis has given us the original images *in terms of differences and similarities between them*. The PCA analysis identified the statistical patterns in the data. Since all the vectors are 4 dimensional, we will get 4 eigenvectors. In practice, we are able to exclude some of the less significant eigenvectors with the recognition still performing well.

To analyze 3d matrices there is a more complex procedure performed with a supervised technique such as PLS. In this case, is important to know *a priori* the reference value or the class relative to a specific spectral trend. Applying the provisional model, elaborated on a dataset, it is possible to attribute a class value or a quantitative prediction to each pixel on the image to be analyzed. This could be already the final result (confusion matrix or SEP) or an image representing, on grey scale or B/W, the classification texture or the quantitative prediction.

Segmentation is the second fundamental step for the Hyperspectral analysis: this operation reduces the complexity discriminating the objects from the background. Many different techniques could apply this operation. In the following paragraphs will be shortly shown two possible methods successfully applied, alternative to statistical threshold: *k*-means clustering and Otsu.

K-means clustering is a procedure to classify a given data set through a certain number of clusters (assume k clusters) fixed *a priori*. Its partitions the observations into k mutually exclusive clusters, and returns a vector of indices (Classes) indicating to which one of the k clusters each observation it has been assigned. Each cluster in the partition is defined by its object member and its centroid, that is the point to which the sum of distances from all objects in that cluster is minimized (by iterative procedure). The result is a set of clusters that are as compact and well-separated as possible.

In order to use algorithms in an unsupervised fashion, a method is needed to estimate the number of texture regions and their parameters. It was first developed a statistical model of the observed image based on the texture model used in segmentation. This overall model is parameterized by the number of textures, the parameters of the individual textures, and the rate at which each texture occurs. Than it is shown that, for a fixed number of textures, approximate local maximum parameter estimation may be performed to alternately re-estimating the texture parameters and repartitioning the clusters associated with each texture. This operation is the k-means algorithm that minimizes the mean square deviation about cluster means.

The data given by x is clustered by the *k*-means method, which aims to create partitions the points into *k* groups such that the sum of squares from points to the assigned cluster centres is minimized. At the minimum, all cluster centres are at the mean of their Voronoi sets (the set of data points which are nearest to the cluster centre).

The Otsu method uses the zero-th and first-order cumulative moments of a grey-value histogram to find the optimal threshold that will separate the two main classes of an image, background and foreground.

Another way of accomplishing similar results is to set the threshold trying to make each cluster as tight as possible, thus minimizing their overlap. Obviously, we can't change the distributions, but we can decide where we separate them (the threshold). As we adjust the threshold one way, we increase the spreading of one and decrease the spreading of the other. The goal then is to select the threshold that minimizes the combined spread.

Infrared thermal imaging equipment

Among the different image analysis techniques and technologies, thermography has the capability to associate to the image information, the thermal punctual information, which is the temperature of each single pixel, in order to operate comparisons between objects inside the same image. Moreover, a good termographic system could record until 50 frames *per* second; in this way is possible to operate dynamic measurements observing temporal modification of the object. For example it is possible to observe the development of thermal gradients within specific areas of the object that could be related to the beginning of usage or disgregative phenomena. This information could also be used to study biological phenomena.

Thermal imaging devices provide the observer with instruments that can collect (just like a video or still camera) and convert the thermal infrared radiation emitted (and reflected) by objects into images that can be seen on a view screen or computer display.

Every object at temperatures above Absolute Zero (0 °K or -273.15 °C) emits thermal radiation, most in the infrared portion of the electro-magnetic spectrum, that is a function of its temperature (T^4) and its emission or emissivity capability (Stefan Boltzman law for real objects).

$$Wb = \varepsilon \, \sigma \, T^4 \tag{4}$$

Many objects that are very hot emit thermal radiation that present in the visible and even in the ultraviolet portion of the EM spectrum as well as the infrared.

For temperature inferior to 500 °C, the maximum of irradiation is observed for wavelengths on middle IR. There are sensors able to detect, with a discrete accuracy, radiation at these wavelengths (*long wave* thermal cameras; 7.5-14 µm). Those systems could calculate the surface temperature of the object, operating corrections due to may factors: physical, environmental and instrumental. However, one of the most important factors for a correct IR thermal measurement is the emissivity that measures the capability of an object to adsorb or emit the thermal radiation. This capability is equal to 1 only for ideal objects (black objects) but is in general very low (< 0.4) for most metals, while is major to 0.85 for plastic or organic materials.

The complement of emissivity is reflectivity that is the capacity to reflect the incident radiation.

The setting of correct value of emissivity depends also by the geometry of the material and the recording angle, for this reason is important in the quantitative thermal survey. Otherwise the most important character that influences the emissivity value is the kind of material and his superficial structure.

Thermal imagers are one of the predictive maintenance tools being widely used in commerce and industry. Thermal imagers with temperature measurement capability, called sometimes 'Radiometric Imagers' and 'Quantitative Thermal Imagers' are used also in many Non-destructive testing situations and some of the professional societies for non-destructive testing around the world, such as the American Society for Non-destructive Testing (ASNT) in USA, have adopted Infrared Testing as a sanctioned testing method.

Advanced fourth generation 320 x 240 uncooled microbolometer FPA detector, which allows obtaining noise-less, high resolution longwave images with 76,000 picture elements. Real-time scanning at standard video rates (60Hz) allows viewing fast moving targets and dynamic events.

As spectral images, also thermal images or thermal videos can be processed by visual inspection or automated procedure. Visual inspection is the basic common type of analysis, especially when no radiometric inference is needed. Other very common analysis is the expression of mean temperature value (or max, min, st. dev) identified by the operator in a specific area (ROI).

Automated image elaboration is often use in scientific analysis with the purpose to distinguish or mapping homogeneous radiometric or temperature areas among the picture. Because of thermoimages expression in 2D matrices, usually no multivariate preprocessing is needed. The most common function adopted is the segmentation or greyscale level reduction through statistical thresholding, k-means, OTSU methods and other.

Only in very sophisticated analysis, multivariate techniques are performed. When the goal is to observe and interpreter dynamic situations (thermal transitory), a time-step sequence of images of the same scene (thermal video) can be acquired. Thermal video can be expressed as 3D matrix, where the Z dimension represents the time trend of radiation or temperature. Multivariate technique, normally PCA, can be applied to reduce amount of information in Z dimension and to rebuild new processed images based on eingevalues.

Application examples

Spectral visual inspection: differences between imaging and punctual spectrometers on beef meat

This test was performed in order to analyze and indicate the possible differences, in a metrological sense, between two different spectral approaches: imaging and punctual spectrometers. Other differences examined in this example concern the presence of environmental light and the background colour. Those variables, often neglected, could definitely influence the measure and so the final results of the analysis.

Experimental protocol:

- For each meat slice was performed 5 acquisitions on 5 different points (Figure 4).
- Using Punctual spectrometer, data were acquired using a reflectance probe β45/0 (Figure 5).
- Using the spectral scanner (Imspector V1000), for each meat slice a spectral image was acquired (Figure 6), and, afterwards, for each of the 5 delimitated areas of the slice, a ROI was manually extracted.

Spectral data was then analyzed with an unsupervised pattern analysis (PCA). Both data from punctual and spectral analysis was used to observe if there were evident differences between external variables: environmental light; white (1) *vs.* black (2) background. Figure 7 show that there are not distinguishable groups. Those external variables do not induce relevant variations.

Afterwards both data from punctual and spectral analysis was used to observe if there were evident differences between external variables: white background; environmental light (1) *vs.* dark (2). Also Figure 8 show that there are not distinguishable groups. Those external variables do not induce relevant variations.

Observing if there were instrumental differences between punctual spectrometer (1) and spectral scanner (2) with black background and environmental light, in Figure 9 it is possible to observe that the two groups are perfectly distinguishable.

Figure 4. Five different point of acquisition on a meat slice.

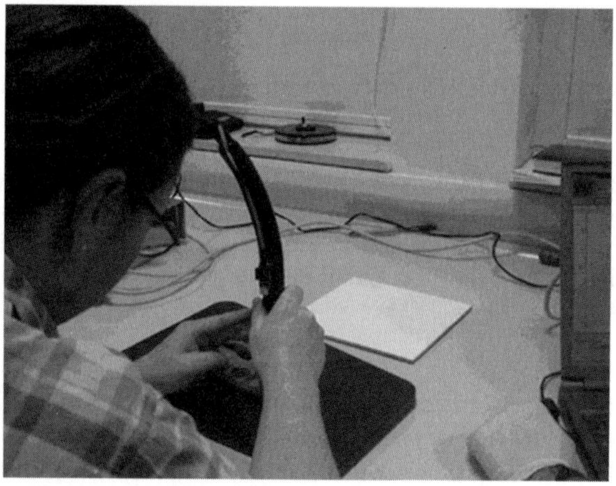

Figure 5. Operator using the punctual spectrometer.

Figure 6. Meat slice on the spectral scanner.

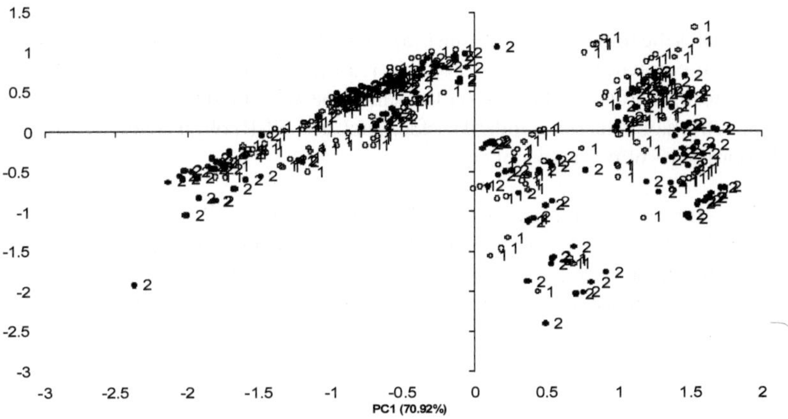

Figure 7. Scatter plot of the first two axes on the PCA on data with environmental light. White background is evidenced as 1, black background as 2.

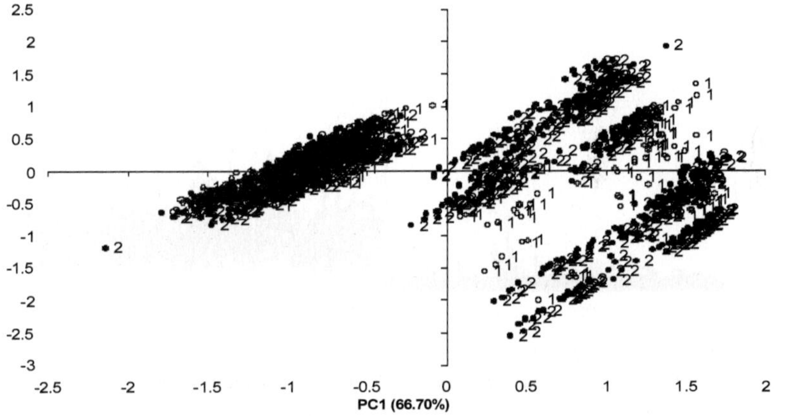

Figure 8. Scatter plot of the first two axes on the PCA on data with white background. Environmental light is evidenced as 1, dark evidenced as 2.

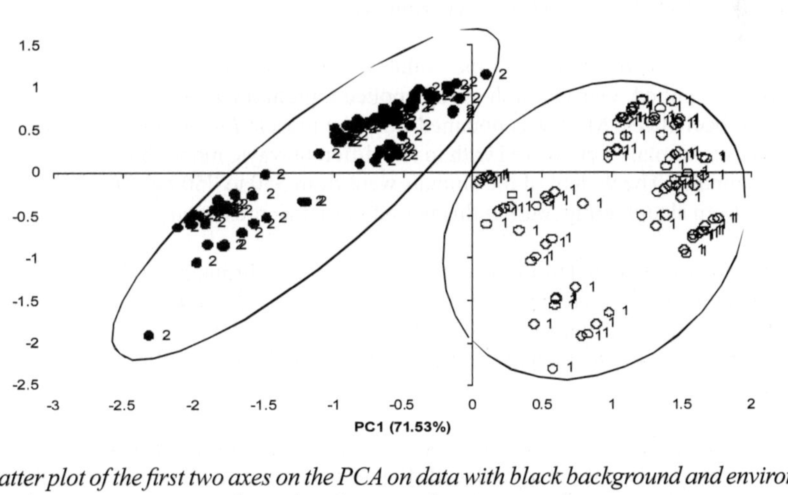

Figure 9. Scatter plot of the first two axes on the PCA on data with black background and environmental light. Punctual spectrometer is evidenced as 1, spectral scanner as 2.

This result indicates the importance of the instrumental effect on spectral determination and, similarly, the difficulties in comparing measures. For this reason in order to eliminate errors, where is not possible to perform a cross calibration on the same spectral standard, is important to indicate the reflectance spectrum of at least a standard surface such as a NLP colour reference (the UK's National Physical Laboratory - http://www.npl.co.uk).

Finally, in the Figure 10, the unexplained residual variance of the PLS model (*Qres*) are reported, in the comparison between the two instruments - Spectral scanner (1) and punctual (2) – on the basis of the measurements on the same samplings. *Qres* represents the information that could be considerate as a noise in the measurements. In this case spectral scanner measurements are much more noisy than punctual spectrometer ones.

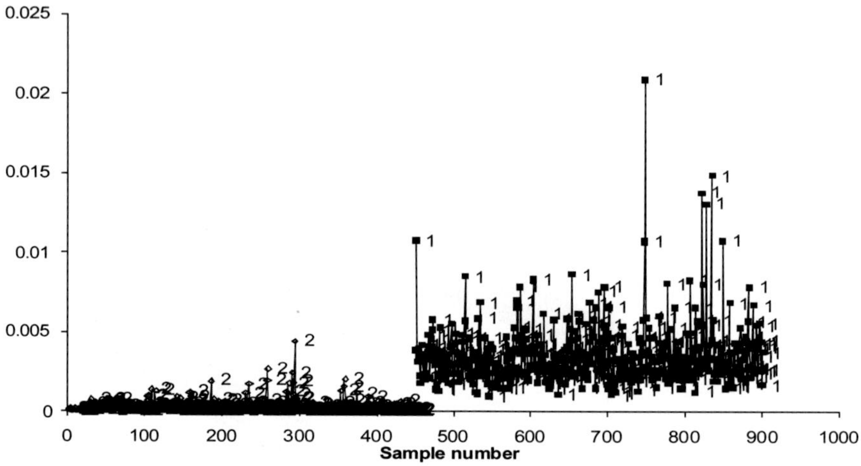

Figure 10. Unexplained residual variance of the PCA model (Qres). image spectral scanner (1) and punctual spectrometer(2).

Hyperspectral imaging for beef meat classification

Hyperspectral imaging techniques for the evaluation of carcass and meat quality in cattle and sheep are not already applied. Our research group applied a preliminary evaluation test performing this technique on beef slices. Meat was obtained from the muscle *Longissimus dorsi* of 12 animals of Chianina cattle. Animals were reared with in two different ways: indoor box (6 animals) and natural pasture (6 animals). The weight of the animals went from 550 to 750 kg. Analyses were performed after 8 days from slaughtering; during this period meat was preserved at 1-4 °C.

The experiment was conducted in order to evaluate if the chromatic and spectral characteristics of meat was influenced by the type of rearing (box, pasture) to allow an effective classification of the rearing type. Spectral images of meat slices were acquired with the spectral scanner VIS-Nir previously described. Acquisition procedure was the same for all samplings (B/W calibration, dark room, reflectance expression).

The first analysis applied was the visual inspection, described above. Mean curves of mean values of reflectance for all pixels in the analyzed ROI (Figure 11) indicate a higher value for the meat relative to the animals breed in box indoor for all the wavelengths. PCA classification applied to means ROI's spectra for each sampling (Figure 12) show a complete distinction between different rearing types.

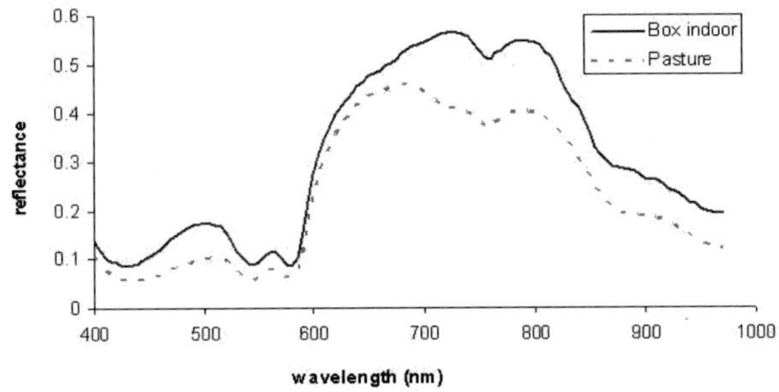

Figure 11. Mean curves of mean values of reflectance for all pixels in the analyzed ROI for Box indoor and Pasture animals.

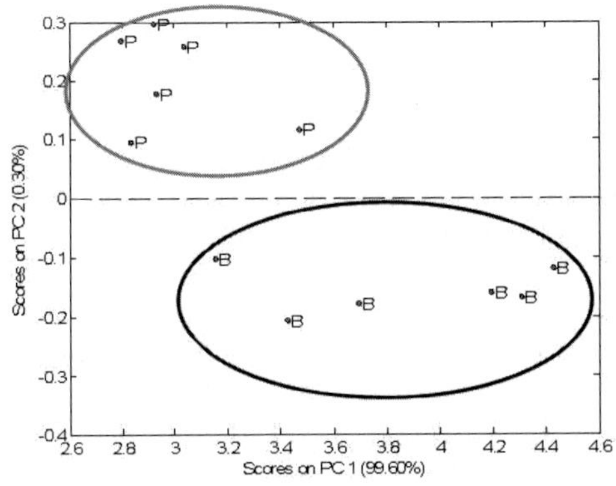

Figure 12. PCA classification applied to means ROI's spectra for each sampling.

Spectral images of these samplings were then analyzed with the following hyperspectral imaging procedure:

1. PCA processing of the 3D matrices and extraction of the first 2 PC's scores for each pixel;
2. reconstruction of 8-bit greyscale images through the distribution of the PC's scores in the interval 0 – 255 levels;
3. segmentation of the PC's images by different methods (Otsu, k-means clustering) and grey-levels (2, 4, 8, 16);
4. extraction of basic textural parameters such as percentage of covered area (white objects) in respect to dark background.

All the analyses were performed in Matlab environment (toolboxes PLS and image analysis).

Two examples, one for each rearing type, of full hyperspectral analysis are reported in Figure 13 and Figure 14. It is possible to notice that the 8-bit image correspondent to PC1 (80% explained variance) is in both cases well representative of meat texture characteristics where areas with a strong grey-level contrast are present. The chemical-physical contribute (Vis or NIR) could indicate a connective tissue, fat or different quantity of water. In the present experiment those information were not available, also for the extreme difficulties in chemically mapping small areas of the meat slice. Then, considering images reconstructed from scores of PC2 (that in both cases represents a very little percentage of the explained total variance) it is possible to notice, for the typology pasture, some interesting secondary structures. Also these structures, to a more in-depth exam, could be related to specific features. For the indoor box typology, the image of PC2 is, instead, completely confused, without a consistent structure, representing practically only noise. This analysis is also completely stretchable to all the other images belonging to the same rearing type.

Observing segmented images it is possible to notice as, only the 2-bits images (Black and White) are sufficiently intelligible and usable for a quantitative analysis of texture. Between the two proposed methods, Otsu and k-means, the first seems to be the most reliable and correctly attributing the chromatic classes (Black and White) to the lighter or darker pixels of the 8-bits image. 4- and 16-bits images obtained from segmentation are certainly richer of chromatic shades and texture, but, for the correct interpretation, the local chemical-physical properties are needed. Image segmentation 'noise' (PC2 rearing type box) produce outputs difficult to interpret, meanwhile in pasture rearing, specific structure are present. The analysis on the image texture conducted on the two rearing type show (Table 1) as the mean percentage of covered area is superior in the indoor box (38%) than in pasture (30%).

The statistical comparison (t-test) is not significant, for the high dispersion of values and the low number of samplings. However, results follow the trend observed in Figure 11, where indoor box reflectance is always superior to pasture. The same comparison on PC2 results, instead, doesn't reveal any differences. Hyperspectral imaging can represent a highly informative methodology in the evaluation of the carcass and meat of sheep and cattle.

Table 1. Mean percentage of covered area (white) in respect to background (black) measured on OTSU 2 bit images on the PC's for the two different rearing systems.

		Box	Pasture
PC1	Average	38.2%	30.8%
	StdDev	14.9%	13.3%
PC2	Average	44.8%	44.3%
	StdDev	3.0%	9.2%

Infrared thermal imaging

The Agritechlab group conduced a preliminary analysis to perform infrared thermal imaging of lamb carcasses to monitoring temperature variation due to the increasing chilling storage time.

For the thermal image acquisition, a FLIR (S40) thermo-camera was used. The principal characteristics are: detector type, Focal plane array (FPA) uncooled microbolometer; Field of view (FOV), 24° - at distance of 1 m the FOV is equal to 0.42 x 0.31 m; Instantaneous field of View (IFOV), 1.3 mrad (the theoretical FOV of one pixel); image frequency, 60 Hz; spectral range, 7.5 to 13 μm, optical resolution 320x240 pixel, focus, automatic or manual; thermal sensitivity @ 50/60Hz, 0.08 °C at 30 °C; temperature range -40÷ +120 °C; accuracy (% of reading) ±2 °C or ±2%.

Emissivity (ε) was set equal to 0.98 because of the acquisition was done on biological materials. Assuming that e=1–reflectivity, the infrared radiation quote acquired by the camera resulted from environmental reflection was very little and therefore not measured.

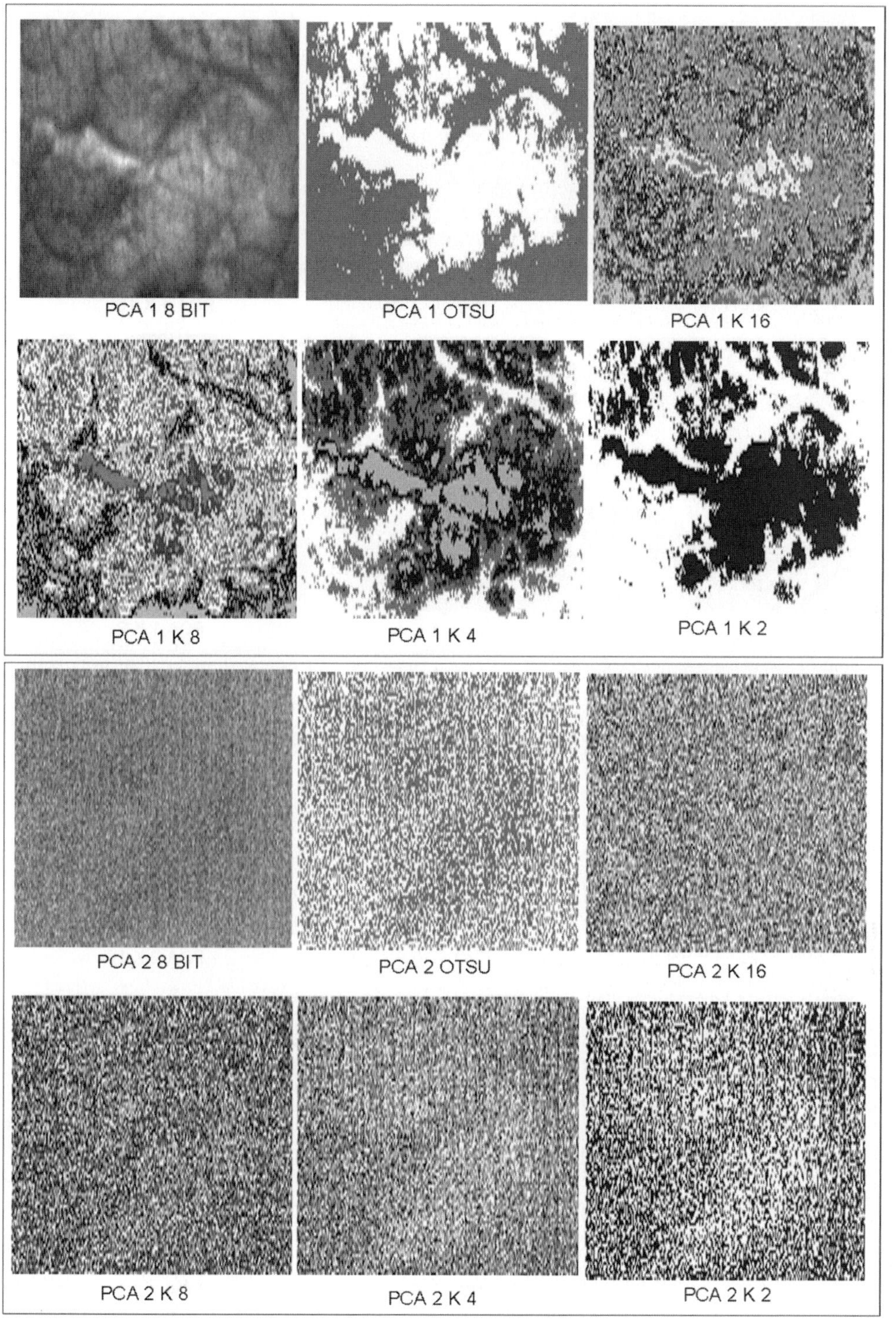

Figure 13. Hyperspectral processed images obtained from one beef spectral image corresponding to rearing type: box.

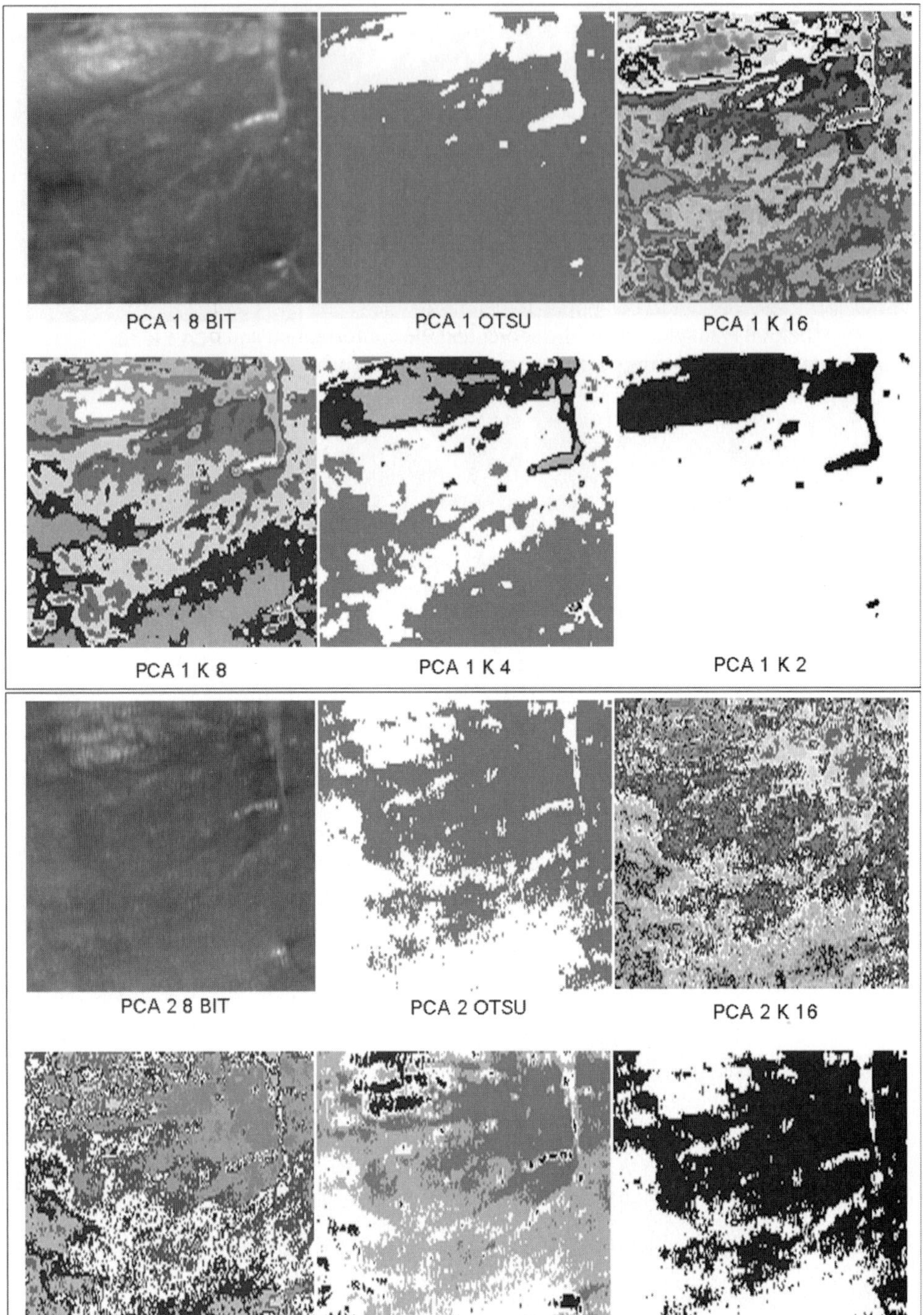

Figure 14. Hyperspectral processed images obtained from one beef spectral image corresponding to rearing type: pasture.

For each trial, a thermal video was acquired at 10 images/s for approximately 3 min, always in the same position.

First trials were conducted to evaluate the capacity of infrared thermovision to monitor temperature progress in different area of lamb carcasses in chilling storage. Basic idea is that low modification of local surface temperature could represent an image of chemo-physical deterioration and of quality changes in the end.

In chilling storage of lamb carcasses, just after slaughter two thermal periods could be distinguished: the fast cooling form ambient to storage temperature and the stationary chilling regime. Infrared imaging thermography could be applied to determine the speed and the uniformity of the process due both to thermal characteristic of the cooler and the conformation and quantity of the carcasses. Thermovision monitoring during chilling storage is expected to give significant information concerning the onset and the evolution of microbiological and biochemical modifications of specific areas of the lamb carcasses, also in relation to the presence of fat, connective or muscle tissues.

As first trial, no specific measurement of chemo-physical parameters was related to the infrared thermal images acquired on a single carcass at different chilling storage days (1, 2, 3, 6, 7, 8). Specific purpose was to verify the technological approach to infrared thermo-vision in monitoring chilling storage of animal carcasses. For infrared imaging acquisition the thermo-camera (Flir S40) above described was used. Different images of total body carcass (day 1, after slaughter) and half side body (days 2, 3, 6, 7, 8) were acquired by a trained operator, directly in the refrigerator, limiting to few minutes his presence at visual distance of 1.5 m.

Offline temperature measurements were performed by trained operator on the stored thermal images in six different spots with regard to the internal part of the half-side lamb. For temperature calculations the emissivity value was setup equal to 0.98. The spots underline the maximum temperature differences between carcass surfaces during chilling storage.

In Figure 15 it can be noticed that inside the body, some areas lose the heat less than other external structures as legs or chest. In Figure 16 the thermal images of the lamb carcass are shown at different chilling storage days. Increasing storage time, determines the development of warmer areas inside the chest even if maximum temperature value is always less than the limit of a good conservation practice (4 °C). Some values measured in specific point on thermal images are reported in Table 2.

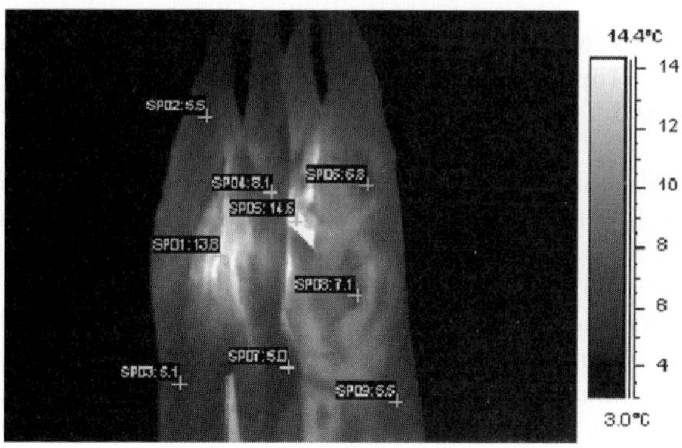

Figure 15. Temperature imaging and spot measurement of lamb carcasses in pre-cooling phase just after slaughter.

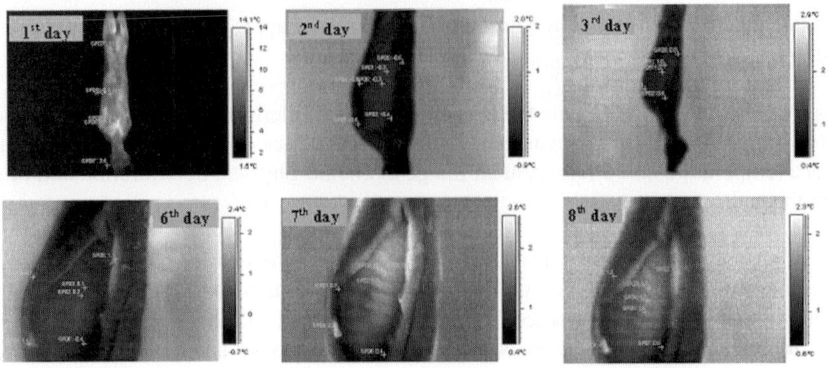

Figure 16. Infrared thermographical imaging of the lamb carcass at different chilling storage days.

Table 2. Progress of temperature values in relation to different days of chilling storage of the lamb carcass.

Storage days (n)	Average T °C	Min T °C	Max T °C
1	10.6	2.6	13.9
2	-0.5	-0.8	-0.2
3	0.9	0.6	1.1
6	0.5	-0.4	1.1
7	1.6	0.4	2.3
8	1.5	0.6	1.9

The increasing value of mean temperature could be related to the longer chilling storage. The results of this first trial can be consider only as a positive indication of the use of infrared thermal imaging in monitoring the thermal conditions of animal carcass during chilling storage.

Conclusions

Novel photonics approaches such as spectral visual inspection, hyperspectral and infrared thermal imaging could allow more sophisticated mapping analysis for carcass grading and meat quality evaluation of sheep and cattle.

There are only few international papers on this subject indicating, for us, a difficult and complex application more than a lower analytical importance of these non-contact and non-destructive techniques.

The fast increasing importance of the associated technology and the improvement of more powerful and sophisticated methods of analysis and processing allow to predict a wider applicative spreading in the next future, until the setup of an expert system able to work online. Our hope is that the short methodological approaches reported in this work, could, however, help and encourages further development and the future widening of the argument.

References

Bjarno, O.C., 1982. Multicomponent Analysis of Meat Products by Infrared Spectroscopy. Journal of the Association of Official Analytical Chemist, Volume 65, 696-700.

Chao, K., Y.R. Chen, W.R. Hruschka and B. Park, 2001. Chicken heart disease characterization by multispectral imaging. Transaction of the American Society of Agricultural and Biological Engineers, Volume 17, 99-106.

Chao, K., P.M. Mehl and Y.R. Chen, 2002. Use of hyper- and multi-spectral imaging for detection of chicken skin tumors. Transaction of the American Society of Agricultural and Biological Engineers,, Volume 18, 113-119.

Chen, Y.R., 1992. Classifying diseased poultry carcasses by visible and near-IR reflectance spectroscopy. Optics in Agriculture and Forestry, Volume 1836, 46-55.

Davis, J.C., 1986. Statistics and Data Analysis in Geology. John Wiley & Sons, New York.

De Jong, S., 1993. SIMPLS: an alternative approach to partial least squares regression. Chemometrics and Intelligent Laboratory Systems, Volume 18, 251-263.

Geesink, G.H., F.H. Schreutelkamp, R. Frankhuizen, H.W. Vedder, N.M. Faber, R.W. Kranen and M.A. Gerritzen, 2003. Prediction of pork quality attributes from near infrared reflectance spectra. Meat Science, Volume 65, 661-668.

Harper, D.A.T., 1999. Numerical Palaeobiology. John Wiley & Sons, Chichester, UK.

Hsieh, C, Y.R. Chen, B.P. Dey and D.E. Chan, 2002. Separating septicemic and normal chicken livers by visible/near-infrared spectroscopy and back-propagation neural networks. Transaction of the American Society of Agricultural and Biological Engineers,, Volume 45, 459-469.

Ibarra, J.G., Y. Tao, A.J. Cardarelli and J. Shultz, 2000. Cooked and raw chicken meat: emissivity in the mid-infrared region, Applied Engineering in Agriculture, Volume 16, 143-148.

Irie, A. and F. Iwaki, 2003. Fibre-optic method for estimation of bovine fat quality. Journal of the Sciences of Food and Agriculture, Volume 83, 483-486.

Jeyamkondan, S., G.A. Kranzler, B.J. Morgan and S. Rust, 2004. Predicting beef tenderness using near-infrared spectroscopy. In: Bennedsen, B.S., Y.R Chen, G.E. Meyer, A.G. Senecal and S.I. Tu (eds.), Proceeding of the International Society for Optical Engineering 'Monitoring Food Safety, Agriculture, and Plant Health', Volume 5271, 356-365.

Josell, A., L. Martinsson, C. Borggaard, J.R. Anderseb and E. Tornberg, 2000. Determination of RN phenotype in pigs at slaugter-line using visual and near-infrared spectroscopy. Meat Science, Volume 55, 273-278.

Kruggel, W.G., R.A. Field, M.L. Riley, H.D. Radloff and K.M. Horton, 1981. Near-infrared reflectance determination of fat, protein and moisture in fresh meat. Journal of the Association of Official Analytical Chemist, Volume 64, 692-696.

Lanza, E., 1983. Determination of moisture, protein, fat and calories in raw pork and beef by near-infrared spectroscopy. Journal of Food Science, Volume 48, 471-474.

Lawrence, K.C., W.R. Windham, B. Park, D.P. Smith and G.H. Poole, 2004. Comparison between visible/NIR spectroscopy and hyperspectral imaging for detecting surface contaminants on poultry carcasses. Proceeding of the International Society for Optical Engineering, Volume 5271, 35-42.

Leroy, B., S. Lambotte, O. Dotreppe, H. Lecocq, L. Istasse and A. Clinquart, 2003. Prediction of technological and organoleptic properties if beef Longissimus thoracis from near-infrared reflectance and transmission spectra. Meat Science, Volume 66, 45-54.

Maldague, X.P.V. (ed.), 1994. Infrared Methodology and Technology. Nondestructive Testing Monographs and Tracts, Volume 7, CRC Press, Boca Raton, FL, USA.

Martens, H., E.A. Bakker, and K.I. Hildrum, 1981. Application of near-infrared reflectance spectrometry in the analysis of meat products. In: Prändl, O. (ed.), Proceedings of 27th European Meeting of Meat Research Workers, Wien, Austria, 516-564.

Menesatti, P., M. Pincu, S. Carzaniga, M. Iacurto, E. Martinelli, F. Stanco and M. Bogliani, 2003. Utilizacion del técnicas opticas para detectar diferencias espectrales relacionadas a diferentes tiempos de conservaciòn de carne envasada al vacìo. VII Congreso Argentino de Ingenierìa Rural, Balcarce, Argentina, n. 2-2.

Menesatti, P., M. Iacurto, S. D'Andrea, M. Pincu and C. Tipaldi, 2004. La termografia all'infrarosso per rilevare in benessere animale. L'Informatore Agrario, Issue 45, 1-4.

Park, B. and Y. R. Chen, 2001. Co-occurrence matrix texture features of multispectral images on poultry carcasses. Journal Agricultural Engineering Researcher, Volume 78, 127-139.

Park, B., K.C. Lawrence, W.R. Windham and D.P. Smith, 2004. Multispectral imaging system for fecal and ingesta detection on poultry carcasses. Journal of Food Process Engineering, Volume 27, 311-327.

Rahkonen, J and H. Jokela, 2003. Infrared radiometry for measuring plant leaf temperature during thermal weed control treatment. Biosystems Engineering, Volume 86, 257-266.

Skjervold, P.O., R.G. Taylor, J.P. Wold, P. Berge, S. Abouelkaram, J. Culioli and E. Dufour, 2003. Development of intrinsic fluorescent multispectral imagery specific for fat, connective tissue, and myofibers in meat. Journal of Food Science, Volume 68, 1161-1168.

Swatland, H. J.,1997. Relationships between the back-scatter of polarised light and the fibre-optic detection of connective tissue fluorescence in beef. Journal of the Sciences of Food and Agriculture, Volume 75, 45-49.

Tao. Y. and J.G. Ibarra, 2001 Far-infrared and 3D imaging for doneness assessment in chicken breast. In: Chen, Y.R. and S.I. Tu (eds.), Proceeding of the International Society for Optical Engineering 'Photonic Detection and Intervention Technologies for Safe Food', Volume 4206, 185-192.

Van den Berg, R.A., H.C.J. Hoefsloot, J.A. Westerhuis, A.K. Smilde and M.J. Van der Werf, 2006. Centering, scaling, and transformations: improving the biological information content of metabolomics data. BMC Genomics, Volume 7, 142.

Yang, C.C., Y.R. Chen and K. Chao, 2005. Development of multispectral image processing algorithms for identification of wholesome, septicemic, and inflammatory process chickens. Journal of Food Engineering, Volume 69, 225-234.

Yang, C.C., K. Chao, Y.R. Chen, M.S. Kim and H.L. Early, 2006. Simple multispectral image analysis for systematically disease chicken identification. Transaction of the American Society of Agricultural and Biological Engineers, Volume 49, 245-257.

Genes affecting meat quality

Alessio Valentini
Department of Animal Productions, Università della Tuscia, Viterbo, Italy

Abstract

Several traits of economical importance in meat production are under genetic control. The search for Quantitative Trait Loci (QTL) has been particularly fruitful as it has allowed to identify several chromosomal regions which are statistically associated with variation of economically important traits in many domestic species, however, in a very few cases the finding of a QTL has allowed to positional cloning the gene(s) of interest, like for Calpastatin, DGAT1, CLPG. Several other genes have been discovered by comparative genomics as it is the case of Myostatin, Leptin and RYR. Very powerful techniques are now in use that will help discovering new genes affecting meat quality in short time. Several livestock species are fully sequenced and a large number of SNP are available to perform QTL search by very fine maps and eventually uncover the causative polymorphisms. Transcriptomic and proteomic analysis permit today to detect under and overregulated genes within of thousands of transcripts and proteins. In this ways genes have been discovered that affect fatty acids proportions, lipid metabolism pathways, cell growth and/or maintenance, protein syntheses as well as cell proliferation. However, very much is expected from the polymorphisms in regions that control genetic expression. Some very promising examples are within 5' regulatory regions in GHR and Myostatin genes, while others are in the 3' UTR that are target of miRNA's and bring to mRNA suppression. As the tools for selection become more powerful, more care should be paid to maintain the genetic variability within the livestock species. In parallel with the selection for the desirable alleles, a survey of the erosion of genetic diversity should be implemented, with actions able to restore it while preserving the economic sustainability.

Keywords: meat quality, QTL, SNP, selection

Introduction

Several traits of economical importance in meat production are under genetic control. Some of them are known and already used in selection schemes. Moreover, several traits, like carcass and meat quality are measured *post-mortem*. Therefore, the identification of genes behind these traits and selection for favourable alleles by testing livestock in vivo is particularly necessary for meat producers. The availability of genetic variants affecting meat traits will help also in rapidly switching selection objectives in a market where consumers and retailers demands change also quickly.

Different strategies for the identification of genes have been employed to dissect complex traits. Namely, (1) QTL search followed by positional cloning and (2) candidate gene analysis followed by validation. Recently, some new promising strategies have been employed which show highly promising results and implications.

QTL approach

The search for Quantitative Trait Loci (QTL) has been particularly fruitful as it has allowed to identify several chromosomal regions which are statistically associated with variation of economically important traits in many domestic species (i.e. Table 1 and Figure 1). This has been achieved using both experimental populations (F2, backcross etc.) and outbred populations exploiting within family linkage disequilibrium (daughter, grand-daughter design etc.). A search in http://www.animalgenome.org in early 2007 gave 253 QTLs for meat quality in pig and 23 in cattle. However, several more can be found in cattle using the 'Bovine QTL Viewer' at http://bovineqtlv2.tamu.edu/index.html (Figure 1)

In a very few cases the finding of a QTL has allowed to positional cloning the gene(s) of interest. This is mainly due to the limited number of markers in the genetic maps used. Only recently very dense maps have been compiled and full genome sequences have been completed and made available in the public domain. Therefore, the confidence limits of the regions harbouring the QTL are yet quite large and span several millions of bases, making it difficult and extremely expensive to identify the constitutional polymorphism(s) causative of the phenotypic variation. In many circumstances, the knowledge of the genes and their function in the region of QTL has been of help in focusing on the most likely causative region.

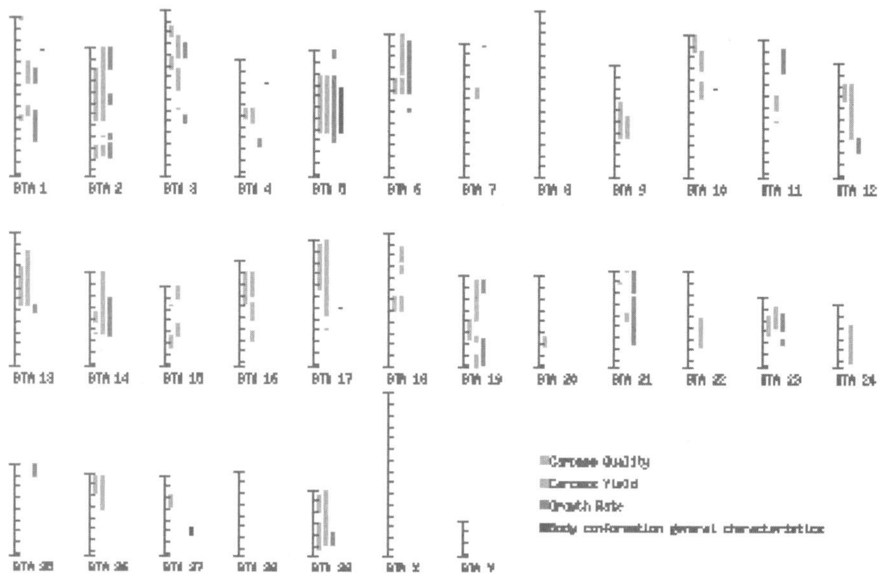

Figure 1. Bovine QTL for meat traits at "Bovine QTL Viewer".

CAST

Suggestive quantitative trait loci (QTL) affecting average Instron force and other tenderness measures were mapped to pig chromosome 2 (Malek *et al.*, 2001) by using a three-generation intercross between Berkshire x Yorkshire (B x Y) pigs. Investigation of the region revealed the presence of Calpastatin (*CAST*) in the region. This is a specific inhibitor of calpains, a Ca^{2+}-activated protease family and considered to be the major cause of initiation of myofibrillar protein degradation in living muscle. Extensive analysis of the *CAST* gene revealed several polymorphisms (Rothschild *et al.*, 2004; Rothschild and Ciobanu, 2004) which altered the protein and these had large effect on tenderness (Table 1).

Table 1. Meat quality parameters according to genotypes at CAST genes (Ciobanu et al.*, 2004).*

Trait [b]	Mean	SD	P^c	Haplotype contrasts [a]		
				1 vs. 2	1 vs. 3	2 vs. 3
Firmness score	3.43	0.63	0.01	–0.23	–0.06	0.17
P				0.006	0.1	0.04
Avg. Instron force, kg	4.36	0.87	0.009	0.07	–0.14	–0.21
P				0.53	0.006	0.06
Sensory tenderness score	7.85	1.15	0.17	0.04	0.14	0.1
P				0.81	0.06	0.54
Sensory chewiness score	2.41	0.92	0.09	–0.1	–0.12	–0.02
P				0.42	0.03	0.87
Sensory juiciness score	6.03	1.47	0.04	0.16	0.22	0.06
P				0.41	0.01	0.74

[a]Haplotype 1: *249Lys-638Arg* (frequency = 0.50); Haplotype 2: *249Arg-638Arg* (0.07); Haplotype 3: *249Arg-638Ser* (0.43).

[b]Number of observations varied from 448 to 482.

[c]*P*-value for overall test of the effect of calpastatin haplotypes.

DGAT1

Grisart *et al.* (2002) used a positional cloning approach to identify a non conservative lysine to alanine substitution (*K232A*) in the bovine *DGAT1* gene that was proposed to be the causative quantitative trait nucleotide underlying a quantitative trait locus (QTL) affecting milk fat composition, previously mapped to the centromeric end of bovine chromosome 14. Afterwards, using a candidate gene approach, Thaller *et al.* (2003) investigated the effects of the same polymorphism of *DGAT1* on the fat content of *musculus semitendinosus* in 55 cattle individuals (28 German Holstein and 27 Charolais). The lysine/alanine polymorphism of *DGAT1* on the fat content of m. *semitendinosus* showed significant effects in both breeds, where the lysine allele showed to be the more efficient version of the enzyme with regard to triglyceride synthesis, being associated with a higher lipid content of different tissues.

CLPG

The trait Callipyge first appeared in a sheep born in Oklahoma in 1983. After the ram's distinct features emerged, he was named Solid Gold and saved from slaughter. Although the hefty haunches initially seemed valuable, breeders have tried to eliminate the mutation because the meat in callipyge animals is not tender due to the high content of calpastatin which is a strong inhibitor of calpains and of the myofibrillar degradation. Two independent efforts to identify the CLPG mutation in sheep have resulted in refinement of physical and comparative maps of the region on chromosome 18 (Fahrenkrug *et al.*, 2000; Berghmans *et al.*, 2001). Animals with genetic recombination events were used to reduce the candidate interval to an approximately 400-Kb region containing plausible candidate genes such as delta, Drosophila, homolog-like (*DLK1*; also known as *PREF-1*) and maternally expressed gene 3 (*MEG3*; also known as *GTL2*). The homologous regions on human chromosome 14 and mouse chromosome 12 have been intensively studied because *DLK1* and *MEG3* are mapped in this zone and are reciprocally imprinted and expressed from the paternal and maternal alleles, respectively (Schmidt *et al.*, 2000).

This region has been fully sequenced, particularly in subjects that were homozygous for markers in the region and heterozygous for the *CLPG* gene and an A to G transition was found causative of the phenotype It was hypothesised that the mutation was recent and these subjects inherited a large block of chromosome without any recombination since the mutation onset, with the exception in the *CLPG locus*. Eventually, the mutation was identified in intron 6 of a putative human gene. Interestingly, none of the candidate genes present in the region were involved in the trait determination. Recently, Takeda *et al.* (2006) demonstrated that the *CLPG* mutation alters the muscular epigenotype of the *DLK1-GTL2* IG region in *cis*, including hypomethylation, acquisition of novel DNase-I hypersensitive sites, and strongly enhanced bidirectional, long-range intergenic transcription.

Candidate gene approach

This approach takes advantage from information gathered from other species, quite often human and model species, or from other production types (e.g. *DGAT1*). Afterwards, the gene is analysed in the species of interest to search for polymorphisms that can be associated with the trait.

RYR/HAL

The best known and well described major gene affecting meat quality in pig is the so called Halothane sensitivity gene (*HAL*), also referred to a porcine stress syndrome (PSS) gene and has been studied and discussed extensively. The PSE defect of meat, characterised by pale, soft and exudative muscles has the same inheritance as PSS or MHS (malignant hyperthermia syndrome). The adverse effects of PSE meat results in meat appearance which is unacceptable for the consumer, lowers the production yield of cooked ham and increases the seasoning loss of dry-cured ham. Malignant hyperthermia also affects carcass traits and it is well documented that halothane positive pigs give heavier, shorter and leaner carcasses than halothane negative pigs. On the whole the gene definitely improves carcass lean content, but has a detrimental effect on pH, colour, drip loss, intramuscular fat and reduces meat tenderness and juiciness.

Pigs that are homozygous for the recessive *HAL* n allele have porcine stress syndrome (PSS) and they are subject to sudden death from stress. In addition, those surviving and those heterozygous for the condition have pale, soft and exudative (PSE) meat caused by the rate and /or the extent of post-mortem pH decrease. The mutation causing PSS and PSE is now known and is within the porcine ryanodine receptor (*RYR1*) that maps to chromosome 6 (Fujii *et al.,* 1991). The *RYR1* gene encodes the skeletal muscle ryanodine receptor, which serves as a calcium release channel of the sarcoplasmic reticulum as well as a bridging structure connecting the sarcoplasmic reticulum and transverse tubule (MacLennan *et al.*, 1989). The associated biochemical alteration is in Ca^{2+} release channel (*CRC1* gene) of the sarcoplasmic reticulum of the muscle cell. A DNA test for the defective allele (HAL1843 ™) is patented and used widely throughout the world. This test allows to distinguish the genotyping of pigs into normal (NN), carriers (Nn), and stress susceptible (nn).

Worldwide, the frequency of this defective allele has decreased to nearly zero, though some lines maintain the gene in order to capture the increased lean produced from heterozygous pigs. Recently an interaction has been proved between *RYR* and Insulin-like growth factor 2 (*IGF2*) gene (*IGF2* intron3-g.3072G>A). *IGF2* expression in *m. longissimus dorsi* of animals with mutations in both *IGF2* and *RYR1* was significantly lower than in animals that inherited the *IGF2* mutation but were homozygous wildtype for *RYR1* (Stinckens *et al.*, 2007).

LEP

The leptin gene (*LEP*) is the homologue of the murine *LEP* gene also known as Obese or Ob. *LEP* encodes a protein composed of 146 amino acids that is expressed in adipose tissue, and that acts as a major regulator for food intake and energy homeostasis. Genetic defects in the murine and human leptin genes are associated with extreme obesity and infertility. Following the identification of the leptin gene by positional cloning in obese mice (Zhang *et al.*, 1994) considerable efforts have been made to understand its role in obesity, adipose deposition, feed intake and energy homoeostasis. The function of leptin has been studied in different species and a positive relationship between a positive energy balance and circulating levels of leptin has been observed (Barb *et al.*, 2001).

The melanocortin-4 receptor has a key role in concert with leptin in regulating feed intake and energy balance. The significant association of *MC4R* mutation to human obesity was found with the identification of non-functional (frame-shift) mutations. More than 30 additional amino acid variations in human *MC4R* have been found in diverse ethnic groups and some of them were associated with genetically obese human subjects and impaired in vitro cellular functions. The mutations in the *MC4R* gene represent the most common genetic cause of non-syndromic human obesity (Yeo *el al.*, 2003; Kim *et al.*, 2004). Therefore, leptin and melanocortin-4 receptor were considered a very plausible candidate gene for many economically important traits also in livestock. In the cattle *LEP* several polymorphisms were discovered by sequencing the various regions of the gene and promoter, and some of them were found associated to carcass fat deposition, feed intake and performance traits (Buchanan *et al.*, 2002; Lagonigro *et al.*, 2003; Barendse *et al.*, 2005; Liefers *et al.*, 2005). Seven polymorphisms were found in the swine leptin gene and the association analyses with production traits have produced disagreeing results depending on the experimental population studied (Jiang *et al.*, 1999; Kennes *et al.*, 2001). Further polymorphisms in LEP gene are under investigation in Casertana, an autochthons Italian swine breed (D'Andrea *et al.*, 2007). In cattle, numerous polymorphism associated with traits, were identified in melanocortin-4 receptor gene and in pro-opiomelanocortin (POMC) gene that is the precursor of alpha melanocyte stimulating hormone (αMSH) an agonist of MC4R (Buchanan *et al.*, 2005). A missense mutation Asp298Asn in the porcine MC4R gene has been reported to be associated with growth, fatness and feed intake traits (Kim *et al.*, 2000; Huston *et al.*, 2004). However, no effects have been detected in other populations (Park *et al.*, 2002). Promising results in both gene have also been achieved studying the different expression of LEP and MC4R genes in Casertana and Large White, two pig breeds with divergent phenotypes (D'Andrea *et al.*, 2005).

Efficiency of identification strategies of genes for genetic improvement of livestock

From the examples so far cited, it appears that there is not a clear cut between the two approaches and that often both are pursued together in order to increase the efficiency of the experimental effort. However, in some cases particular strategies proved to be more efficient in pinpointing the causative genes behind an economical important trait.

Myostatin

'Double muscling' (DM) is a trait present in several cattle breeds. DM carcasses generally have a higher dressing percentage, a larger proportion of muscle and lower proportion of fat and bone (Arthur, 1995), making this trait economically interesting. The DM trait is almost fixed in some breeds like Belgian Blue and Piemontese while in other breeds, like Marchigiana, the causative mutation is present at low frequency (Marchitelli *et al.*, 2003).

In the last half of the nineties, a big experiment was carried out to map the trait in an experimental population (Charlier *et al.*, 1995). The gene was mapped in a short region of *BTA2* close to *TGLA44* microsatellite marker (Figure 2). Comparative map in human revealed *COL3a* as a possible candidate gene within the region (Figure 3).

While work was progressing towards positional cloning of the gene and possibly investigating the *COL3* gene polymorphisms, McPherron *et al.* (1997) very rapidly discovered that i. in the mouse the GDF8 gene gives rise to hyperplasic/hypetrophic phenotype when knocked out; ii. a 11 bp deletion in the homologous gene in Belgian Blue cattle could be responsible of DM phenotype because of a disruption in the final gene product (McPherron and Lee, 1997). Once the gene was recognized, several other mutations were identified with different contribution in determining the DM phenotype, like a cistein to tyrosin substitution in the Piemontese (Kambadur *et al.*, 1997) that makes difficult to form a dimer, and the appearance of a stop codon in the third exon that produces a truncated protein (Marchitelli *et al.*, 2003).

Therefore, in this particular case, the use of a model species, which can be easily manipulated and for which a large amount of knowledge of genomic features is available, led to a rapid discovery of the implicated gene, its characteristics, the role in the metabolic pathways and the potential economic use.

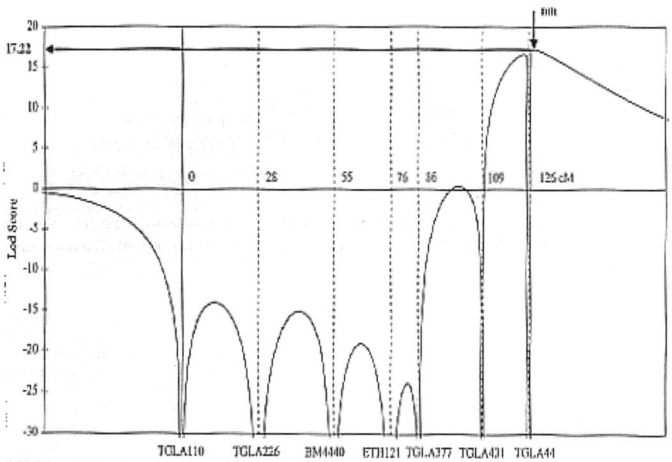

Figure 2. Probability of QTL for DM in BTA2.

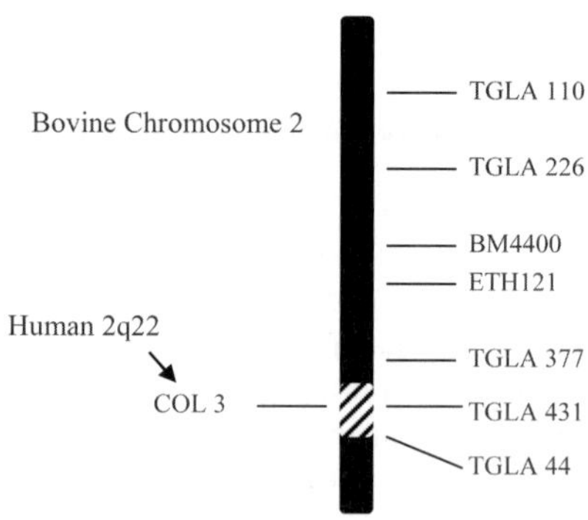

Figure 3. Markers and candidate gene in the region where DM maps.

New methods for the identification of genes and selection

Fine mapping of QTL/genes by high throughput SNP platforms

One by-product of full genome sequencing is the availability of a high number of SNP with a limited additional cost. This is obtained simply by sequencing the genome of a few (hopefully diverse) individuals. The SNP found (usually in the 10,000-100,000 range) can be assayed on high throughput platforms producing very dense maps with strong linkage disequilibrium within adjacent markers. For example, the Illumina system is based on a fibre-optic assembly of 96 individual arrays each array can hold 1,536 different oligo probes (with defined Illumicodes i.e. signatures of genes). Therefore, in a few hours 96 individuals may be genotyped at tens of thousands of *loci*.

The Affymetrix GeneChip® Bovine Mapping 10K SNP Kit contains approximately 10,000 SNPs that are multiplexed in a single assay. Approximately 92 percent of the SNPs for this project were discovered by the Bovine Genome Sequencing Project. The first draft was placed in the public domain for use by biomedical and agricultural researchers, worldwide, in October 2004. The remaining eight percent were obtained from CSIRO (Australia's Commonwealth Scientific and Industrial Research Organisation) and beef cattle cross breeds. All SNPs in this panel have been validated in 84 unique bovine samples from six strains of cattle (Angus, Brahmin, Hereford, Holstein, Jersey, and Limousin). Currently (2007) several cattle breeds are characterised for some thousands of SNPs under the bovine HapMap project.

SNP information can be gathered using *in silico* facilities like the Interactive Bovine In Silico SNP (IBISS) database that has been created at the Commonwealth Scientific and Industrial Research Organisation (CSIRO) to create an *in silico* SNP database by harnessing the vast number of bovine EST sequences available in the public domain (Hawken *et al.*, 2004). In cattle there is already available a panel of about 20,000 SNP. The pig sequencing consortium has recently published the first draft sequences and it is likely that a similar SNP panel will be available shortly.

Microarray and proteomic analysis

Eisen *et al.* (1998) were among the first to use microarray gene expression profiles for gene function prediction. They used hierarchical clustering algorithms to analyze time-course gene expression profiles of budding yeast and human fibroblasts and found that grouping of transcripts with similar profiles across a time dimension could accurately identify genes of similar function. Today microarrays used very sophisticated statistical procedures that account for the multiplicity of contrasts and are helpful to identify the genes involved in a particular production or physiological status, or the metabolic pathway involved.

Microarrays have been used to compare breeds with different attitudes. Wang *et al.* (2005) found that genes more highly expressed in Japanese Black (JB) versus Holstein are associated with unsaturated fatty acid synthesis, fat deposition, and the thyroid hormone pathway. Their results are consistent with the increased amounts and proportions of monounsaturated fatty acids observed in the muscle of JB animals. Lin and Hsu (2005) compared gene transcription profiles of LM between two pig breeds, Duroc and Taoyuan. The strongly upregulated genes in Duroc pigs included those encoding the complex of myofibrillar proteins (e.g., myosin light and heavy chains, and troponin), ribosomal proteins, transcription regulatory proteins (e.g., skeletal muscle LIM protein 1 (SLIM1) and high-mobility group proteins), and energy metabolic enzymes (e.g. electron-transferring flavo-protein dehydrogenase, NADH dehydrogenase, malate dehydrogenase, and ATP synthases). However, the Authors used a human based microarray, therefore the outcomes should be carefully and thoroughly validated by other means (e.g. rtPCR). Pig specific microarrays however exist and they are largely employed: Ponsuksili *et al.* (2007) compared liver expression profiles of pigs of two breeds, the obese German Landrace (DL) and the lean Pietrain (Pi). They found upregulation of key genes of lipid metabolism pathways (*FASN*, *ACSS2*, *ACACA*) in obese DL pigs and upregulation of genes of cell growth and/or maintenance, protein syntheses as well as cell proliferation pathways (*PPARD*, *POU1F1*, *IGF2R*) in lean Pi pigs.

Proteomics, i.e. the study of several hundreds or thousands of proteins differentially present in two or more theses, begins to be employed in meat producing animals. While rather expensive and more complex than microarray technique, proteomics takes into consideration also post-translational modifications and in this way is more related to the phenotype. Bouley *et al.* (2005) used differential proteomic analysis of *Semitendinosus* muscle using two-dimensional gel electrophoresis followed by mass spectrometry. They showed that thirteen proteins, corresponding to 28 protein spots, were significantly altered in response to the myostatin deletion in Belgian Blue The observed changes in protein expression were consistent with an increased fast muscle phenotype, suggesting that myostatin negatively controls mainly fast-twitch glycolytic fibres number.

Expression Quantitative Trait Loci (eQTL)

Brem *et al.* (2002) carried out genetic linkage analysis of genome-wide expression patterns in a cross between a laboratory strain and a wild strain of *Saccharomyces cerevisiae*. They used microarray data (differential intensity of fluorophores) as phenotypes for genetic mapping in a segregating yeast sample. This approach was followed by the mapping of expression quantitative trait loci (eQTL) in mammalian species. Each study identified *cis* and *trans* traits. A *cis* trait is one that genetically maps to the physical location of the gene encoding its mRNA, suggesting that variation at the locus is responsible for the heritable changes in gene expression. A *trans* trait maps to a region distinct from its physical location and thus implies the location of a potential regulator acting in *trans*. The experiments may take advantage by the combined use of Gene Ontology (GO) of correlated transcripts, as they are hypothesized to be regulated by the same transcription factors (Lan *et al.*, 2006). No eQTL studies have been published so far in livestock species, but there is a growing literature particularly on mice where experimental mapping populations are quite easy to construct.

Selection using polymorphisms in regulatory regions

Most of the genes so far cited present alleles that produce proteins with different structures and therefore affect the phenotype by altering a 'normal' behaviour (with the exception of the Callipyge gene). As an example, in the particular case of myostatin, all mutations leading to the double muscling phenotype give rise to a truncated protein or one that cannot form the normal functional dimer. However, there is a huge potential for selection of alleles affecting the regulatory regions of RNA transcription. In this way the protein produced is the normal variant and only its amount is modulated. Several regions of the gene affect the efficiency of transcription, like the 5' promoter, enhancers, silencers, etc.

GHR

The single and combined effects of polymorphisms in the 5'-noncoding region of the bovine growth hormone receptor (*GHR*) gene was examined on traits related to feed intake and meat production in beef cattle. Maj *et al.* (2004) studied four different single nucleotide polymorphisms (SNP). Seventy-one young bulls belonging to four beef breeds (Charolais, Limousin, Aberdeen Angus, Hereford) and one dual purpose breed (Simmental) were included in the study. The results showed that genetic variants at the 5'-noncoding region of the bovine *GHR* gene had a marked effect on beef production traits. In addition, statistically significant associations were shown between combined *GHR* genotypes and feed consumption, carcass weight and dimensions.

GDF8

Crisà *et al.* (2003) found that individuals of Marchigiana beef breed carrying a TT genotype in a polymorphic site of the 5' region of myostatin gene, when associated to the heterozygous status at the III exon, have a muscularity index 25% higher than individuals with all the other possible combinations (Table 2).

Interestingly this SNP exists also in pig and the mutation has been associated with growth traits, showing that individuals with TA genotype had a higher average daily gain than those with TT genotype (Jiang *et al.*, 2002).

The *GDF8* allele of Texel sheep is characterized by a G to A transition in the 3' UTR that creates a target site for mir1 and mir206, microRNAs (miRNAs) that are highly expressed in skeletal muscle. This causes translational inhibition of the myostatin gene and hence contributes to the muscular hypertrophy of Texel sheep (Clop *et al.*, 2006). Analysis of SNP databases for humans and mice demonstrates that mutations creating or destroying putative miRNA target sites are abundant and might be important effectors of phenotypic variation.

Table 2. Combined effect on muscularity index of polymorphisms in promoter region (DRA I) and in III exon of myostatin (Crisà et al.*, 2003).*

III exon	Dra I	Muscularity value	Std. Err.
Wild [a]	TA	109.7	5.51
Wild [a]	TT	106.2	3.97
mh/+ [a]	TA	105.9	11.41
mh/+ [b]	TT	130.6	5.9

mh = mutated allele.
Different superscript letters mean P<0.05.

Conclusions

Several genes have been discovered in livestock species that significantly affect meat traits. From past experience we can better address the strategies for discovering economically important genes. New methods, like high throughput SNP genotyping, eQTL and exploitation of regulatory regions polymorphism, will rapidly lead to the discovery of several useful genetic variants. However, the facility to select for a specific variant of economic interest may bring livestock populations to very fragile conditions. Even now, mostly with traditional systems of genetic evaluation the effective number of population (Ne) for livestock breeds has dramatically reduced. For example, Ne is as low as 46 in French Holstein, a breed that counts 2.5 million animals across France. In Germany, similar results were found with a Ne of 52. An even more extreme results was found in Japan, where the Japanese Black cattle had a Ne of 17.2 in between 1993 and 1997, despite a census size of 0.53 million reproductive cows (from Taberlet *et al.*, 2007). In parallel with the selection for the desirable alleles, a survey of the erosion of genetic diversity should be implemented, with actions able to restore it while preserving the economic sustainability.

References

Arthur, P.F., 1995. Double muscling in cattle: a review. Australian Journal of Agricultural Research, Volume 46, 1493-1515.

Barb, C.R., J.H. Hausman and K.L. Hoseknechtm, 2001. Biology of leptin in the pig. Domestic Animal Endocrinology, Volume 21, 297-317.

Barendse, W., R.J. Bunch and B.E. Harrison, 2005. The leptin C73T missense mutation is not associated with marbling and fatness traits in a large gene mapping experiment in Australian cattle. Animal Genetics, Volume 36, 71-93.

Berghmans, S., K. Segers, T. Shay, M. Georges, N. Cockett and C. Charlier, 2001. Breakpoint mapping positions the callipyge gene within a 285 kilobase chromosome segment containing the GTL-2 gene. Mammalian Genome, Volume 12, 183-185.

Brem, R.B., G. Yvert, R. Clinton and L. Kruglyak, 2002. Genetic dissection of transcriptional regulation in budding yeast. Science, Volume 296, 752-755.

Bouley, J., B. Meunier, C. Chambon, S. De Smet, J.F. Hocquette and B. Picard, 2005. Proteomic analysis of bovine skeletal muscle hypertrophy. Proteomics, Volume 5, 490-500.

Buchanan, F.C., C.J. Fitzsimmons, A.G. Van Kessel, T.D. Thue, D.C. Winkelman-Sim and S.M. Schmutz, 2002. Association of a missense mutation in the bovine leptin gene with carcass fat content and leptin mRNA levels. Genetic Selection Evolution, Volume 34, 105-116.

Buchanan, F.C., T.D. Thue, P. Yu and D.C. Winkelman-Sim, 2005. Single nucleotide polymorphisms in the corticotrophin-releasing hormone and pro-opiomelancortin genes are associated with growth and carcass yield in beef cattle. Animal Genetics, Volume 36, 127-131.

Charlier, C., W. Coppieters, F. Farnir, L. Grobet, P.L. Leroy, C. Michaux, M. Mni, A. Schwers, P. Vanmanshoven, R. Hanset, M. Georges, 1995. The MH gene causing double-muscling in cattle maps to bovine chromosome 2. Mammalian Genome, Volume 6, 788-792.

Clop, A., F. Marcq, H. Takeda, D. Pirottin, X. Tordoir, B. Bibe, J. Bouix, F. Caiment, J.M. Elsen, F. Eychenne, C. Larzul, E. Laville, F. Meish, D. Milenkovic, J. Tobin, C. Charlier and M. Georges, 2006. A mutation creating a potential illegitimate microRNA target site in the myostatin gene affects muscularity in sheep. Nature Genetics, Volume 38, 813-818.

Ciobanu, D.C., J.W. Bastiaansen, S.M. Lonergan, H. Thomsen, J.C. Dekkers, G.S. Plastow and M.F. Rothschild, 2004. New alleles in calpastatin gene are associated with meat quality traits in pigs. Journal of Animal Science, Volume 82, 2829-2839.

Crisà, A., C. Marchitelli, M.C. Savarese and A. Valentini, 2003. Sequence analysis of myostatin promoter in cattle. Cytogenetic and Genome Research, Volume 102, 48-52.

D'Andrea, M., M. Fidotti and F. Pilla, 2005. Differences in MC4R mRNA levels between Casertana and Large White pig breeds. Italian Journal of Animal Science, Volume 4, Suppl. 2, 94-96.

D'Andrea, M., F. Pilla, D. Waddington and A. Archibald, 2007. Lack of association between 67 and 1 polymorphisms in LEP and MC4R genes respectively, and traits in Casertana, Large White and Cross Breed swine. Manuscript in preparation.

Eisen, M.B., P.T. Spellman, P.O. Brown and D. Botstein, 1998. Cluster analysis and display of genome-wide expression patterns. Proceeding of the National Academic of Science of the United States of America, Volume 95, 14863-14868.

Fahrenkrug, S.C., B.A. Freking, C.A. Rexroad III., K.A. Leymaster, S.M. Kappes and T.P.L. Smith, 2000. Comparative mapping of the CLPG locus. Mammalian Genome, Volume 11, 871-876.

Fujii, J., K. Otsu, F. Zorzato, S. De Leon, V.K. Khanna, J.E. Weiler, P.J. O'Brien and D.H. McLennan, 1991. Identification of a mutation in the porcine ryanodine receptor that is associated with malignant hypertemia. Science, Volume 253, 448-451.

Grisart, B., F. Farnir, L. Karim, N. Cambisano, J.-J. Kim, A. Kvasz, M. Mni, P. Simon, J.-M. Frère, W. Coppieters and M. Georges, 2004. Genetic and functional confirmation of the causality of the DGAT1 K232A quantitative trait nucleotide in affecting milk yield and composition. Proceedings of the National Academy of Sciences of the United States of America, Volume 101, 2398.

Hawken, R.J., W.C. Barris, S.M. McWilliam and B.P. Dalrymple, 2004. An interactive bovine in silico SNP database (IBISS). Mammalian Genome, Volume 15, 819-827.

Huston, R.D., N.D. Cameron and K.A. Rance, 2004. A melanocortin-4 receptor (MC4R) polymorphism is associated with performance traits in divergently selected large white pig populations. Animal Genetics, Volume 35, 386-390.

Jiang, Z.H. and J.P. Gibson, 1999. Genetics polymorphism in the leptin gene and their association with fatness in four pig breeds. Mammalian Genome, Volume 10, 191-193.

Jiang, Y.L., N. Li, L.X. Du and C.X. Wu, 2002. Relationship of T->A mutation in the promoter region of myostatin gene with growth traits in swine. Yi Chuan Xue Bao [Acta Genetica Sinica], Volume 29, 413-416.

Kambadur, R., M. Sharma, T.P.L. Smith and J.J. Bass, 1997. Mutations in myostatin (GDF-8) in double muscled Belgian Blue and Piedmontese cattle. Genome Research, Volume 7, 910-915.

Kennes, Y.M., B.D. Murphy, F. Pothier and M.F. Palin, 2001. Characterization of swine leptin (LEP) polymorphisms and their association with production traits. Animal Genetics, Volume 32, 215-218.

Kim, K.S., N. Larsen, T. Short, G. Plastow and M.F. Rothschild, 2000. A missense variant of porcine melanocortin-4 receptor (MC4R) gene is associated with fatness, growth, and feed intake traits. Mammalian Genome, Volume 11, 131-135.

Kim, K.S., J.M. Reecy, W.H. Hsu, L.L. Anderson and M.F. Rothschild, 2004. Functional and phylogenetic analyses of a melanocortin-4 receptor mutation in domestic pigs. Domestic Animal Endocrinology, Volume 26, 75-86.

Lagonigro, R., P. Wiener, F. Pilla, J.A. Woolluams and J.L. Williams, 2003. A mutation in coding region of the bovine leptin gene associated with feed intake. Animal Genetics, Volume 34, 371-374.

Lan, H., M. Chen, J.B. Flowers, B.S. Yandell, D.S. Stapleton, C.M. Mata, E.T.-K. Mui, M.T. Flowers, K.L. Schueler, K.F. Manly, R.W. Williams, C. Kendziorski and A.D. Attie, 2006. Combined Expression Trait Correlations and Expression Quantitative Trait Locus Mapping. PLoS Genetic, Volume 2, e6.

Lin, C.S. and C.W. Hsu, 2005. Differentially transcribed genes in skeletal muscle of Duroc and Taoyuan pigs. Journal of Animal Science, Volume 83, 2075-2086.

Liefers, S.C., R.F. Veerkamp, M.F.W. te Pas, C. Delavaud, Y. Chilliard, M. Platje, T. van der Lende, 2005. Leptin promoter mutations affect leptin levels and performance traits in dairy cows. Animal Genetics, Volume 36, 111-118.

Maj, A., J. Oprzadek, A. Oprzadek, E. Dymnicki and L. Zwierzchowski, 2004. Polymorphism in the 5'-noncoding region of the bovine growth hormone receptor gene and its association with meat production traits in cattle Animal Research, Volume 53, 503-514.

Malek, M., J.C.M. Dekkers, H.K. Lee, T.J. Baas, K. Prusa, E. Huff-Lonergan and M.F. Rothschild, 2001. A molecular genome scan analysis to identify chromosomal regions influencing economic traits in the pig. II. Meat and muscle composition. Mammalian Genome, Volume 12, 637-645.

Marchitelli, C., M.C. Savarese, A. Crisà, A. Nardone, P. Ajmone Marsan and A. Valentini, 2003. Double muscling in Marchigiana beef breed is caused by a stop codon in the third exon of myostatin gene. Mammalian Genome, Volume 14, 392-395.

MacLennan, D.H., F. Zorzato, J. Fujii, K. Otsu, M. Phillips, F.A. Lai, G. Meissner, N.M. Green, H.F. Willard, B.A. Britt, R.G. Worton and R.G. Korneluk, 1989. Cloning and localization of the human calcium release channel (ryanodine receptor) gene to the proximal long arm (cen-q13.2) of human chromosome 19. American Journal of Human Genetics, Volume 45, A205.

McPherron, A.C., A.M. Lawer and S.J. Lee, 1997. Regulation of skeletal muscle mass in mice by a new TGF-β superfamily member. Nature, Volume 387, 83-90.

McPherron, A.C. and S.J. Lee, 1997. Double muscling in cattle due to mutations in the myostatin gene. Proceeding National Academy of Science of the United States of America, Volume 94, 12457-12461.

Park, H.B., Ö. Carlborg, S. Marklund and L. Andersson, 2002. Melanocortin-4 receptor (MC4R) genotypes have no major effect on fatness in a Large-White X Wild Boar intercross. Animal Genetics, Volume 33, 155-157.

Ponsuksili, S., E. Murani, C. Walz, M. Schwerin and K. Wimmers, 2007. Pre- and postnatal hepatic gene expression profiles of two pig breeds differing in body composition: insight into pathways of metabolic regulation. Physiol. Genomics, Jan 30 (Epub ahead of print).

Rothschild, M.F. and D.C. Ciobanu, 2004. Novel calpastatin (CAST) alleles. United States Patent Application 20040048267.

Rothschild, M.F., J.P. Bidanel, D.C. Ciobanu, 2004. Genome analysis of QTL for muscle tissue development and meat quality. In: te Pas, M.F.W., M.E. Everts and H.P. Haagsman (eds.), Muscle development of livestock animals. Physiology, Genetics and meat quality. CABI Publishing, 247-266.

Schmidt, J.V., P.G. Matteson, B.K. Jones, X.J. Guan and S.M. Tilghman, 2000. The Dlk1 and Gtl2 genes are linked and reciprocally imprinted. Genes Dev., Volume 14, 1997-2002.

Stinckens A., K. Van den Maagdenberg, T. Luyten, M. Georges, S. De Smet and N. Buys, 2007. The RYR1 g.1843C>T mutation is associated with the effect of the IGF2 intron3-g.3072G>A mutation on muscle hypertrophy. Animal Genetics, Volume 38, 67-71.

Taberlet, P., P. Ajmone-Marsan, A. Valentini, H.R. Rezaei, S. Naderi, F. Pompanon, R. Negrini, Econogene Consortium, 2007. Are cattle, sheep, and goats endangered species? Proceeding Symposium Evolutionary Change in Human-altered Environments: An International Summit to Translate Science into Policy, Los Angeles, USA.

Takeda, H, F. Caiment, M. Smit, S. Hiard, X. Tordoir, N. Cockett, M. Georges and C. Charlier, 2006. The callipyge mutation enhances bidirectional long-range DLK1-GTL2 intergenic transcription in cis. Proceeding National Academy of Science of the United States of America, Volume 103, 8119-8124.

Thaller, G., C. Kühn, A. Winter, G. Ewald, O. Bellmann, J. Wegner, H. Zühlke and R. Fries, 2003. DGAT1, a new positional and functional candidate gene for intramuscular fat deposition in cattle. Animal Genetics, Volume 34, 354-357.

Yeo, G.S., E.J. Lank, I.S. Farooqi, J. Keogh, B.G. Challis and S. O'Rahilly, 2003. Mutations in human melanocortin-4 receptor gene associated with severe familial obesity disrupts receptor function through multiple molecular mechanism. Human Molecular Genetics, Volume 12, 561-574.

Wang, Y.H., K.A. Byrne, A. Reverter, G.S. Harper, M. Taniguchi, S.M. McWilliam, H. Mannen, K. Oyama and S.A. Lehnert, 2005. Transcriptional profiling of skeletal muscle tissue from two breeds of cattle. Mammalian Genome, Volume 16, 201-210.

Zhang, Y., R. Proenca, M. Maffel, M. Barone, L. Leopold and J.M. Friedman, 1994. Positional cloning of the mouse obese gene and its human homologue. Nature, Volume 372, 425-432.

Plasma leptin and insulin-like growth factor I (IGF-I) as potentially phenotypic markers for carcass composition and growth rate in lambs

Eberhard von Borell[1], H. Sauerwein[2] and M. Altmann[1]
[1]Institute of Agricultural and Nutritional Sciences, Martin-Luther-University Halle-Wittenberg, 06108 Halle, Adam-Kuckhoff-Str. 35, Germany
[2]Institute of Animal Sciences, Physiology and Hygiene, Bonn University, 53115 Bonn, Katzenburgweg 7-9, Germany

Abstract

The utilisation of hormones as early predictors for final growth rate and carcass composition in breeding programmes is of particular interest. The adipocyte derived hormone leptin correlates well to the extent of body fat tissue. However, the relationship between leptin concentration in early life and final performance is poorly understood. The aim of this paper was to determine the relationship of leptin and IGF-I during growth to final daily gain and carcass composition in lambs with special emphasis to size and cellularity of the different depots body fat depots. Thirty intact male ad libitum fed lambs were blood sampled at 20, 25, 30, 35, and 40 kg live weight. After slaughtering at 40 kg, lean and the visceral, subcutaneous and intermuscular fat were measured by dissection. The intramuscular fat was analysed in a sample of *M. longissimus* by n-hexane extraction. The fat cell diameter was determined in subcutaneous and perirenal fat. Final daily gain correlated to leptin at 30 and 35 live weight (-0.56 and -0.61, $P < 0.01$) and to IGF-I at 20 to 30 kg live weight ($r = 0.52$ to 0.59, $P < 0.01$). Subcutaneous and intermuscular fat correlated to leptin only if measured just before slaughtering ($r = 0.45$ and 0.36, $P < 0.05$) while the visceral fat showed significant correlations to the leptin concentrations analysed at 25 to 40 kg live weight ($r = 0.41$, $P < 0.05$ to $r = 0.58$, $P < 0.01$). Among the visceral fat depots, omental fat expressed the highest correlations to leptin at 35 and 40 kg live weight ($r = 0.63$ and 0.60, $P < 0.001$). Additionally, leptin concentrations in this weight range increased with increasing fat cell diameters ($r = 0.38$, $P < 0.05$ to $r = 0.59$, $P < 0.001$). The relationship of IGF-I to carcass composition was only significant at 30 kg live weight for lean ($r = -0.42$, $P < 0.05$), intermuscular fat ($r = 0.40$, $P < 0.05$), and intramuscular fat (-0.42, $P < 0.05$). This study indicates that leptin concentration measured at the slaughter weight range has the greatest potential to assess body fat content, whereas an earlier prediction does not seem to be feasible. On the other hand, IGF-I has the potential to serve as an early predictor for final growth performance.

Keywords: lamb, leptin, IGF-I, growth, carcass composition

Introduction

The evaluation of carcass merit in vivo for breeding purposes by subjective judgments or ultrasound measurements are two common methods. On the other hand, the utilisation of hormones as a predictor for growth rate and carcass composition is discussed since several years. Studies have mainly focused on IGF-I especially as an early predictor for carcass composition, growth rate and feed intake in cattle (Davis and Bishop, 1994; Davis and Simmen, 2000; Johnston *et al.*, 2002), lambs (Cameron, 1992; Mears, 1995; Wylie *et al.*, 1997) and pigs (Bunter *et al.*, 2002; Cameron *et al.*, 2003; Te Pas *et al.*, 2003). The profitability of an early assay based selection in the meat industry depends on region-specific breeding and marketing conditions. A model calculation for the selection of beef cattle in Australia showed that an early selection for IGF-I at weaning increased the profit from the base of 15.2 to 16.3% (Wood *et al.*, 2004). The IGF-I test is commercially used since 2004 in Australia's cattle performance test (Graser, 2004).

Leptin is another potential hormone for an assay based prediction of growth or carcass composition. It is released mainly from the adipose tissue and correlates with body fat. Geary *et al.* (2003) recommend leptin as an additional index for fat content in beef. The accuracy for the estimation of carcass fat in lambs by plasma leptin concentrations at the time of slaughter was similar to ultrasound fat thickness measurements (Altmann *et al.*, 2005). Studies on the relationship between leptin measured in an early growth phase and the final carcass composition are scarce (Altmann and Von Borell, 2007). Kawakita *et al.* (2001) considered the correlation between the early leptin concentration and the final backfat thickness as too low for a prediction of adiposity in steers.

The aim of the present paper was to determine the relationship of leptin and IGF-I plasma concentrations during growth to final daily gain and carcass composition of lambs with special emphasis on size and cellularity of several fat depots.

Materials and methods

Animals

Intact male East Frisian Milk sheep (n = 15) and Blackheaded Mutton x East Frisian Milk sheep (n = 15) lambs were used in this study. The lambs were born in late winter and were naturally reared by their dams. They were group-housed in deep litter pens and had *ad libitum* access to concentrates (10.2 MJ NE/kg) and hay. The lambs were weighed weekly. The blood-sampling period began at 20.0 (S.E.M. 0.18) kg and ended at 40.5 kg (S.E.M. 0.23) which corresponded to a mean age of 38.5 (S.E.M. 1.7) and 97.1 (S.E.M. 3.1) days, respectively. Weaning occurred on two fixed days when lambs averaged 60 days of age. The mean live weight at weaning was 27.6 kg. When the lambs reached the final weight of 40 kg they were feed deprived for 24 hours immediately after the last blood collection to reduce the gut fill before slaughter. Perirenal, pelvic, intestinal, and omental fat was weighed and summarised as visceral fat tissue. The left carcass side was dissected into lean, subcutaneous and intermuscular fat and bones. The tissue masses were multiplied by 2 for the calculation of the total carcass composition. The intramuscular fat content was analysed in a sample of the M. longissimus between the 13th and 14th rib by n-hexane extraction (Soxtherm 406, Gerhardt Apparate GmbH & Co. KG, Bonn, Germany). Samples of subcutaneous and perirenal fat were collected and frozen at -20 °C for the determination of the cell diameters.

Blood sampling and hormone assays

Blood samples were drawn at 20, 25, 30, 35, and 40 kg live weight. Five blood samples were collected from each lamb between 9.00 h and 11.00 h in 30 min intervals and then centrifuged. In order to obtain a mean hormone concentration over two hours, sera were pooled for each animal, aliquoted and stored at -20 °C until analysed. Hormone concentrations in pooled samples have been shown to adequately represent those obtained from serial samples (Tiwary, 1986).

Plasma leptin concentrations were analysed with a specific enzyme immunoassay that has been validated for use in several ungulate species, including ovine samples (Sauerwein *et al.*, 2004). The intra-assay coefficient of variation was 6.3%, the inter-assay coefficient of variation was 13.9%, and the limit of detection was 0.3 ng/ml.

Using acid-ethanol extraction and cryo-precipitation as described by Blum and Breier (1994), an enzyme immunoassay was employed for IGF-I analysis. Human IGF-1 (IBT Reutlingen, Germany) was used for standards. The primary antibody was a well-characterised antiserum (#878/4, 20), the secondary antibody was sheep antirabbit-IgG (Sigma, Germany). 100 µl of plasma samples or standard IGF-1 were incubated together with 50 µl of the antiserum for 4 h at 4 °C. Then, 50 µl biotinylated IGF-1 (IBT Reutlingen, Germany) was added. After an overnight incubation at 4 °C, the plates were washed three times (Tween 80, 0.5%), 200 µl Streptavidin-POD per well was added and incubated for 1 h at 4 °C. The plates were washed four times and 150 µl tetramethylbenzidine substrate was added. After incubation for 50 min at room temperature, the dye production was stopped by 50 µl 2 M H_2SO_4 and measured. The intra- and inter-assay coefficients of variation were 12.4% and 15.5%, respectively. The limit of detection was 50 ng/ml.

Measurement of fat cell diameters

The samples were sectioned in 20 µm thick slices on a Microtome-Cryostat and fixed in formol calcium for 5 minutes. The slices were stained with haemalaun and fat was removed by xylen for 30-60 sec. The diameter was measured in 300 cells for each sample using a light microscope with a 4 x 10 objective (Labophot-2, NIKON, Japan) connected to a 3CCD colour video camera (DXC 9100P, SONY, Japan). Fat cell diameters were analysed with an image analysis programme (Lucia-G, version 4.8, NIKON, Japan).

Data analysis

Data from purebred and crossbred lambs were cumulated for the statistical analysis because genotype did not influence hormone concentrations and performance traits. Hormone concentrations in lambs of different live weight were analysed by ANOVA and Duncan's multiple comparison post hoc tests. The relationships between final daily gain, carcass composition and hormone concentrations were described by Pearson correlations and regression equations.

Results

Lambs in this study realised a high level of final daily gain (Table 1) from birth to slaughter which is comparable to performance test results for East Frisian Milk sheep in Germany. Despite the similar live weight at slaughter, the mass of various fat tissues showed relatively high coefficients of variation (19.0 to 39.6%). In contrast, the variation of lean mass was low.

Table 1. Final daily gain and carcass composition of experimental lambs.

	Mean	SEM	CV (%)
Final daily gain, g/day	373	9.4	13.9
Total dissectible fat, g	3,547	129.2	19.9
Visceral fat, g	912	33.7	20.2
Intestinal fat, g	344	11.9	19.0
Omental fat, g	344	16.5	26.3
Perirenal fat, g	193	9.9	28.1
Pelvic fat, g	31	2.2	39.6
Carcass fat, g	2,635	106.7	22.2
Intermuscular fat, g	1,773	67.5	20.9
Subcutaneous fat, g	862	42.3	26.8
Intramuscular fat, %	1.6	0.1	32.7
Lean (g)	12,033	155.4	7.1
Perirenal fat cell diameter (µm)	57.0	1.2	11.6
Subcutaneous fat cell diameter (µm)	66.1	1.7	13.8

Leptin concentrations in plasma increased during the growth period and reached the highest level at 40 kg live weight (Figure 1). The effect of time was significant (P < 0.001). IGF-I levels did not change significantly (P = 0.45) during the growth period (Figure 2). No correlations between leptin and IGF-I were found.

The negative relationship between leptin and final daily gain was strongest at 30 and 35 kg live weight (Table 2). Leptin correlated to the visceral fat at all growth periods except for 20 kg live weight, while the correlations to carcass fat were significant only at 40 kg live weight. Among the visceral fat depots, pelvic fat showed the lowest correlations (P > 0.05) to leptin concentrations. The correlations to omental and perirenal fat tended to be stronger as live weight increased.

IGF-I between 20 and 30 kg live weight was positively correlated to final daily gain (Table 3). A significant relationship to lean and some fat tissues was obvious only at 30 kg live weight.

The correlations between fat cell size and plasma leptin measured during growth are represented in Table 4. Leptin was correlated more strongly to the fat cell diameter in perirenal fat than to the subcutaneous fat. While no relationships could be found between fat cell diameter and leptin measured in lambs of 20 -30 live weight, the leptin concentrations at 35 and 40 kg live weight correlated to the mean fat cell diameter in a range of r = 0.38 to 0.59 (P < 0.05 to 0.001).

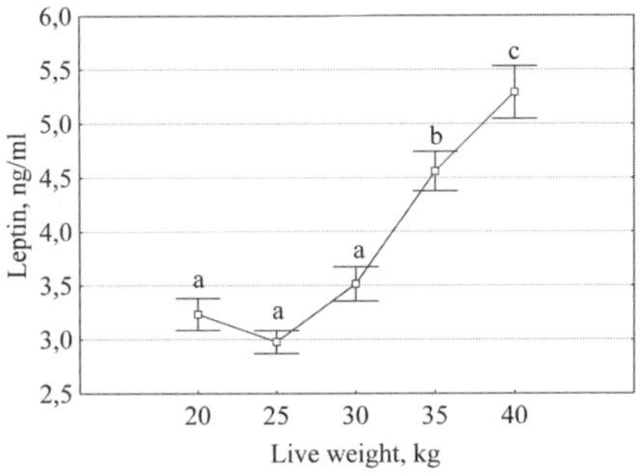

Figure 1. Leptin concentrations during the growth period of 20 to 40 kg live weight (means and standard error of means, means with different letters differ significantly, P < 0.05).

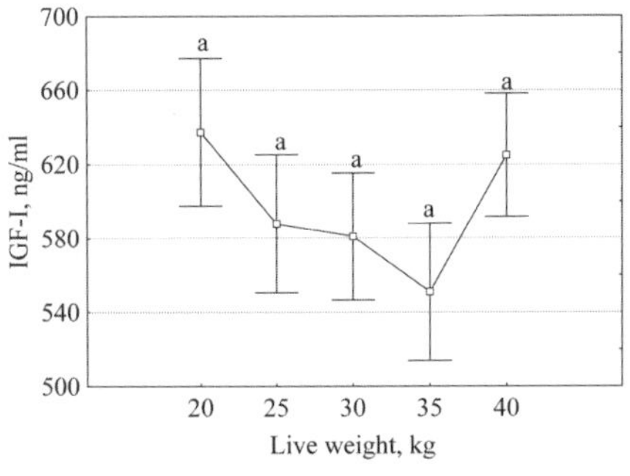

Figure 2. IGF-I concentrations during the growth period of 20 to 40 kg live weight (means and standard error of means, means with different letters differ significantly, P < 0.05).

Table 2. Correlations between leptin measured at different live weights and final daily gain and tissue masses.

	Growth period (kg live weight)				
	20	25	30	35	40
Final daily gain	0.21	-0.10	-0.56**	-0.61**	-0.19
Total dissectible fat	0.09	0.26	0.22	0.26	0.49**
Visceral fat	0.06	0.41*	0.43*	0.54**	0.58**
Intestinal fat	0.09	0.36*	0.43*	0.27	0.40*
Omental fat	-0.03	0.35*	0.38*	0.63***	0.60***
Perirenal fat	0.09	0.36*	0.29	0.46**	0.46**
Pelvic fat	0.16	0.10	0.02	0.06	0.11
Carcass fat	0.09	0.19	0.13	0.14	0.41*
Intermuscular fat	0.14	0.18	0.09	0.08	0.36*
Subcutaneous fat	0.01	0.18	0.19	0.23	0.45*
Intramuscular fat	-0.15	0.06	0.32	0.31	0.19
Lean	0.23	-0.19	-0.34	0.00	0.04

*$P < 0.05$, **$P < 0.01$, ***$P < 0.001$.

Table 3. Correlations between IGF-I measured at different live weights and final daily gain and tissue masses.

	Growth period (kg live weight)				
	20	25	30	35	40
Final daily gain	0.59**	0.52**	0.58**	0.22	0.23
Total dissectible fat	0.08	0.04	0.29	0.23	-0.08
Visceral fat	-0.10	-0.21	0.00	0.14	-0.31
Intestinal fat	0.13	-0.13	-0.04	0.17	-0.23
Omental fat	-0.17	-0.28	-0.13	0.04	-0.28
Perirenal fat	-0.21	-0.12	0.15	0.16	-0.27
Pelvic fat	0.01	0.16	0.42*	0.16	-0.14
Carcass fat	0.13	0.12	0.36*	0.23	0.00
Intermuscular fat	0.17	0.16	0.40*	0.23	-0.03
Subcutaneous fat	0.06	0.05	0.27	0.22	0.05
Intramuscular fat	-0.20	-0.19	-0.40*	-0.05	-0.28
Lean	-0.13	-0.19	-0.42*	-0.06	0.09

*$P < 0.05$, **$P < 0.01$, ***$P < 0.001$.

The correlations in Table 2 and 3 indicate, that final daily gain could be best predicted by IGF-I concentrations at 20 kg live weight and total dissectible fat by leptin concentrations at 40 kg live weight. The regression equations are represented in Figure 3 and 4. The R^2 values for final daily gain and total dissectible fat were 0.35 and 0.24; respectively.

Table 4. Correlations between leptin concentrations during growth and fat cell diameters of subcutaneous and perirenal fat.

Growth period (kg live weight)	Perirenal fat	Subcutaneous fat
20	-0.08	-0.05
25	0.16	0.01
30	0.25	-0.02
35	0.59***	0.38*
40	0.57***	0.49**

*$P < 0.05$, **$P < 0.01$, ***$P < 0.001$.

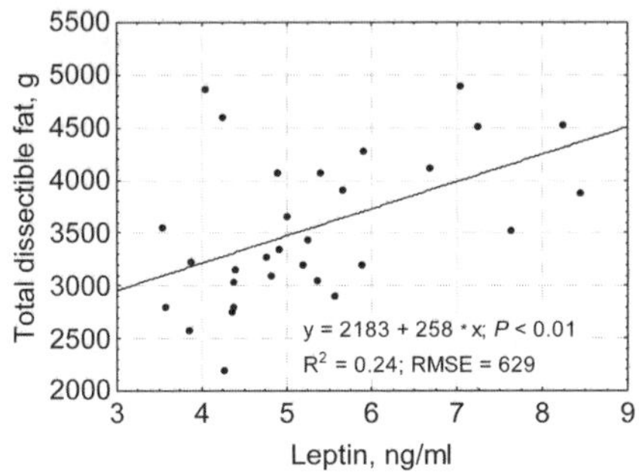

Figure 3. Relationship between leptin concentration at 40 kg live weight and total dissectible fat.

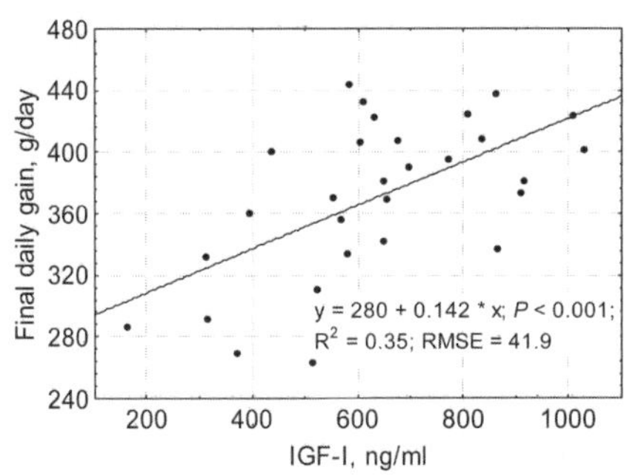

Figure 4. Relationship between IGF-I concentration at 20 kg live weight and final daily gain.

Discussion

The present study was focused on the value of leptin and IGF-I as predictors for final performance traits in growing lambs and examined the relationships of final daily gain and body composition to hormone concentrations measured at several times before slaughtering. The study started at 20 and ended at 40 kg live weight. This growth phase corresponds to the length of lamb performance test in Germany and 40 kg is a common live weight for slaughter lambs in some European countries. A hormonal indicator should be applicable at 20 kg live weight or earlier for a preselection or at 40 kg live weight for the prediction of carcass value in lambs kept for breeding purposes and the carcass composition can not be directly determined by slaughtering.

Leptin

Plasma leptin did not change significantly between 20 and 30 kg live weight within the range in which weaning occurred. However, it increased at 35 kg until time of slaughter. It is assumed that this elevation was caused by a progressive increase of fat accretion. Leymaster and Jenkins (1993) found an accelerated deposition of kidney-pelvic and carcass fat with advanced live weight corresponding with the weight range of this study.

To our knowledge, the relationship of early plasma leptin concentrations to daily gain and carcass composition at slaughter in lambs has not yet been studied. The negative correlation to final daily gain at 30 and 35 live weight implies that slow growing lambs had high leptin concentrations. This can be explained by the inhibitory effect of leptin on feed intake and the stimulatory influence on energy expenditure (Clarke *et al.*, 2001, Henry *et al.*, 1999, Morrison *et al.*, 2001, Tokuda *et al.*, 2000). The correlations to final daily gain at lower live weights (20 and 25 kg) were inconsistent and thus leptin can not be regarded as a suitable predictor of early selection for growth rate. That goes for the prediction of final carcass traits, too. In steers, leptin was also a weak early indicator to predict final carcass traits (Kawakita *et al.*, 2001).

A great number of studies in several species indicated that body fat content is highly associated with blood leptin concentration measured at the same time like body fat. A correlation of 0.68 was found in ewes that varied widely in the deposition of fat tissue (Delavaud *et al.*, 2000). We further observed correlations from 0.40 to 0.64 between blood leptin concentrations and the mass of several fat depots in lambs of 35 and 45 kg live weight showing a lower variation in body fat (Altmann *et al.*, 2005). Similar correlations between plasma leptin measured just before slaughtering at 40 kg live weight and the mass of fat tissues were obtained in the present study.

Plasma leptin concentration is influenced by the amount of fat reserves and the expression of leptin in the adipocytes. The mRNA level was seen to be higher in a fat selection line of sheep than in a lean line (Kumar *et al.*, 1998). Some studies in humans indicated that leptin mRNA and leptin secretion is higher in subcutaneous than in visceral fat (30-33). High correlations were found between plasma leptin and subcutaneous fat mass; however, low or no correlations to the visceral fat mass were observed (Ramachandran *et al.*, 1997; Langendonk *et al.*, 1998; Banerji *et al.*, 1999; Minocci *et al.*, 2000). In contrast to humans, leptin expression in subcutaneous fat is reported to be lower or equal in visceral (perirenal or omental) fat from cattle (Ji *et al.*, 1998; Kim *et al.*, 2000; Ren *et al.*, 2002; Yang *et al.*, 2003) and sheep (Kumar *et al.*, 1998). Leptin mRNA correlated positively to the adipocyte volume in humans (Oberkofler *et al.*, 1997), rats (Villafuerte *et al.*, 2000), mice (Guo *et al.*, 2004) and steers (Yang *et al.*, 2003). Because fat cell diameter in subcutaneous fat of our study was higher than in perirenal fat and the mass of subcutaneous fat was 4-fold higher than the perirenal fat mass, we expected a stronger correlation of plasma leptin to the subcutaneous fat mass than to the perirenal fat. However, this was not observed. Omental fat contributed to 9.7% of total fat, but it contributed with the highest proportion of 36% to the variation of plasma leptin. This indicates a high expression of leptin in omental fat. Another study in steers showed no difference in correlations between leptin and visceral, subcutaneous, and intermuscular fat (Yamada *et al.*, 2003).

The correlations between leptin and body composition indicate that the leptin concentration at the end of the fattening period (40 kg live weight) is the most suitable predictor of body composition. It explained 24% of the variance in total dissectible fat. Similar results were obtained by Altmann *et al.* (2005) for the prediction of carcass fat by leptin concentrations measured before slaughtering in lambs of 35 and 45 kg live weight.

IGF-I

After an early postnatal increase in lambs up to 2-4 weeks of age (Breier and Gluckmann, 1991; Gatford *et al.*, 1997), IGF-I decreased (Suttie *et al.*, 1993; Mears, 1995; Wylie *et al.*, 1997) or remained at the same level (Medrano and Bradford, 1991; Breier and Gluckmann, 1991). In this study, slight but not significant decreasing IGF-I concentrations in lambs between 20 and 35 kg live weight were observed. The majority of circulating IGF-I (75–85%) derives from the liver, at least in mice (Sjögren *et al.*, 1999, Svensson *et al.*, 2005). IGF-I is also expressed in adipose tissue in a depot specific manner and in relation to the adipocyte volume in rats (Villafuerte *et al.*, 2000). Although the IGF-I mRNA levels were as high as in the liver (Peter *et al.*, 1993), they did not correlate to the plasma concentrations of IGF-I. The IGF-I mRNA in ovine skeletal muscles and type I IGF receptors decreased with increasing age (Oldham *et al.*, 1996). The fact that adipose tissue, fat cell size and muscle tissue increases during growth and the lack of a significant age effect on plasma IGF-I in lambs of our study indicates that the contribution of the adipocyte and muscle derived IGF-I to the circulating pool was limited.

Some studies showed that IGF-I concentrations were progressively less correlated with final live weight or final daily gain at successive sampling dates (Medrano and Bradford, 1991; Wylie *et al.*, 1997). Likewise, we observed in this study a strong relationship between IGF-I at 20 to 30 kg live weight and final daily gain whereas the IGF-I concentrations measured in heavier lambs were less associated to final daily gain. Based on these results, we suggest that IGF-I has a potential as an early predictor for growth potential. The IGF-I concentrations measured at 20 kg live weight explained 35% of the variance in final daily gain.

Lambs with a high lean content in carcass had greater IGF-I concentrations than lambs with a high fat content (Cameron, 1992; Francis *et al.*, 1995; Gatford *et al.*, 1996; Wylie *et al.*, 1997). In contrast, Suttie *et al.* (1993), McCann *et al.* (1997), and Francis *et al.* (1998) reported lower IGF-I concentrations in lambs of a lean selection line. They assumed that more IGF-I is bound by tissue receptors in these lambs or that the tissue availability for IGF-I is greater due to different amounts of binding proteins. IGF-I and carcass composition in our study were only correlated at 30 kg live weight for some traits. However, this period is too late for an early preselection of lambs.

Implications

Our data suggest that plasma leptin can not be recommended as an early predictor of total fat. However, it could be used as a second stage selection criterion in breeding rams at 40 kg - the common weight range of slaughter lambs in some European countries. IGF-I measured in an early growth phase (before the beginning of performance test) could be utilised for the prediction of final daily gain. IGF-I is not suitable for the prediction of carcass composition neither at an early age nor in the slaughter weight range. Further studies should clarify if these results are applicable to other breeds or species.

References

Altmann, M., H. Sauerwein and E. Von Borell, 2005. Relationship between plasma leptin concentrations and carcass composition in fattening mutton: a comparison with ultrasound results. Journal of Animal Physiology and Animal Nutrition, Volume 89, 326-330.

Altmann, M. and E. Von Borell, 2007. Review Article: Leptin as an indicator for carcass composition in farm animals. Animal Science Journal, Volume 78, (in press).

Banerji, M.A., N. Faridi, R. Atluri, R.L. Chaiken and H.E. Lebovitz, 1999. Body composition, visceral fat, leptin and insulin resistance in Asian Indian men. Journal of Clinical Endocrinology, Volume 184, 137-144.

Blum, W.F. and B.H. Breier, 1994. Radioimmunoassays for IGFs and IGFBPs. Growth Regulation, Volume 4 (Suppl. 1), 11-19.

Breier, B.H. and P.D. Gluckman, 1991. The regulation of postnatal growth: nutritional influences on endocrine pathways and function of the somatotrophic axis. Livestock Production Science, Volume 27, 77-94.

Bunter, K.L., S. Hermesch, B.G. Luxford, K. Lahti and E. Sutcliffe, 2002. IGF-I concentration measured in juvenile pigs provides information for breeding programs: a mini review. Proceeding 7th World Congress Genetics Applied to Livestock Production, Montpellier, France, Communication No.03-09.

Cameron, N.D., 1992. Correlated physiological responses to selection for carcass lean content in sheep. Livestock Production Science, Volume 30, 53-68.

Cameron, N.D., E. McCullough, K. Troup and J.C. Penman, 2003. Serum insulin-like growth factor-1 concentration in pigs divergently selected for daily food intake or lean growth rate. Journal of Animal Breeding and Genetics, Volume 120, 228-236.

Clarke, I.J., A.J. Tilbrook, A.I. Turner, B.W. Doughton and J.W. Goding, 2001. Sex, fat and the tilt of the earth: effects of sex and season on the feeding response to centrally administered leptin in sheep. Endocrinology, Volume 142, 2725-2727.

Davis, M.E. and M.D. Bishop, 1994. A note on consequences of single-trait selection for insulin-like growth factor 1 in beef heifers. Animal Production, Volume 59, 315-320.

Davis, M.E. and R.C.M. Simmen, 2000. Genetic parameter estimates for serum insulin-like growth factor-I concentration and carcass traits in Angus beef cattle. Journal of Animal Science, Volume 78, 2305-2313.

Delavaud, C., F. Bocquier, Y. Chillard, D.H. Keisler, A. Gertler and G. Kann, 2000. Plasma leptin determination in ruminants: effect of nutritional status and body fatness on plasma leptin concentration assessed by a specific RIA in sheep. Journal of Endocrinology, Volume 165, 519-526.

Francis, S.M., B.A. Veenvliet, S.K. Stuart, R.P. Littlejohn and J.M. Suttie, 1995. Insulin-like growth factor (IGF-I) mRNA and plasma concentrations in lean and fat genotypes of sheep. Proceeding Endocrine Society of Australia, Volume 38, 168.

Francis, S.M., B.A. Veenvliet, S.K. Stuart, R.P. Littlejohn and J.M. Suttie, 1998. Growth hormone secretion and pituitary gland weight in suckling lambs from genetically lean and fat sheep. New Zealand Journal of Agricultural Research, Volume 41, 387-393.

Gatford, K.L., T.P. Fletcher, I.J. Clarke, P.C. Owens, K.J. Quinn, P.E. Walton, P.A. Grant, B.J. Hosking, A.R. Egan and E.N. Ponnampalam, 1996. Sexual dimorphism of circulating somatotropin, insulin-like growth factor I and II, insulin-like growth factor binding proteins, and insulin: relationships to growth rate and carcass characteristics in growing lambs. Journal of Animal Science, Volume 74, 1314-1325.

Gatford, K.L., K.J. Quinn, P.E. Walton, P.A. Grant, B.J. Hosking, A.R. Eganand P.C. Owens, 1997. Ontogenic and nutritional changes in circulating insulin-like growth factor (IGF)-I, IGF-II and IGF-binding proteins in growing ewe and ram lambs. Journal of Endocrinology, Volume 155, 47-54.

Geary, T.W., E.L. McFadin, M.D. MacNeil, E.E. Grings, R.E. Short, R.N. Funston and D.H. Keisler, 2003. Leptin as a predictor of carcass composition in beef cattle. Journal of Animal Science, Volume 81, 1-8.

Graser, H., 2004. IGF-I blood test improves NFI EBVs. Breedplan News, Volume 14, 12.

Guo, K.Y., P. Halo, R.L. Leibel and Y. Zhang, 2004. Effects of obesity on the relationship of leptin mRNA expression and adipocyte size in anatomically distinct fat depots in mice. American Journal of Physiology - Regulatory Integrative and Comparative Physiology, Volume 287, 56-61.

Henry, B.A., J.W. Goding, W.S. Alexander, A.J. Tilbrook, B.J. Canny, F. Duneshea, A. Rao, A. Mansell and I.J. Clarke, 1999. Central administration of leptin to ovariectomized ewes inhibits food intake without affecting the secretion of hormones from the pituitary gland: evidence for a dissociation of effects on appetite and neuroendocrine function. Endocrinology, Volume 140, 1175-1182.

Ji, S., G.M. Willis, R.R. Scott and M.E. Spurlock, 1998. Partial cloning and expression of the bovine leptin gene. Animal Biotechnology, Volume 9, 1-14.

Johnston, D.J., R.M. Herd, M.J. Kadel, H.U. Graser, P.F. Arthur, J.A. Archer, 2002. Evidence of IGF-I as a genetic indicator of feed efficiency traits in beef cattle. 7th World Congress Genetics Applied to Livestock Production, Montpellier, France, Volume 31, 257-260.

Kawakita, Y., H. Abe, K. Hodate, A. Iguchi, M. Kobayashi, T. Mori, K. Kasai, Y. Tanai, Y. Kanbe, H. Mashiyama, T. Asada and T. Miyashige, 2001. The relation between plasma leptin concentrations and carcass lipid contents in Japanese Black steers. Livestock Production Science, Volume 73, 25-34.

Kim, H., Y. Chi, K. Chung, K. Kim, Y. Choi and M. Baik, 2000. Differential response of obese gene expression from fasting in bovine adipose tissues. Bioscience Biotechnology and Biochemistry, Volume 64, 2240-2242.

Kumar, B., S.M. Francis, J.M. Suttie and M.P. Thompson, 1998. Expression of obese mRNA in genetically lean and fat selection lines of sheep. Comparative Biochemistry and Physiology - B Biochemistry and Molecular Biology, Volume 120, 543-548.

Langendonk, J.G., H. Pijl, A.C. Toornvliet, J. Burggraaf, M. Frolich, R.C. Schoemaker, J. Doornbos, A.F. Cohen and A.E. Meinders, 1998. Circadian rhythm of plasma leptin levels in upper and lower body obese woman: influence of body fat distribution and weight loss. Journal of Clinical Endocrinology and Metabolism, Volume 83, 1706-1712.

Leymaster, K.A. and T.G. Jenkins, 1993. Comparison of Texel- and Suffolk-sired crossbred lambs for survival, growth and compositional trait. Journal of Animal Science, Volume 71, 859-869.

McCann, J.P., S.C. Loo, D.L. Aalseth and T. Abribat, 1997. Differential effects of growth hormone stimulation on fasting and prandial metabolism and plasma insulin-like growth factors (IGFs) and IGF-binding proteins (IGFBPs) in lean and obese sheep. Journal of Endocrinology, Volume 154, 329-346.

Mears, G.J., 1995. The relationship of plasma somatomedin (IGF-I) to lamb growth rate. Can. Journal of Animal Science, Volume 75, 327-331.

Medrano, J.F. and G.E. Bradford, 1991. Growth performance and plasma insulin-like growth factor I concentrations in sheep selected for high weaning weight. Journal of Animal Science, Volume 69, 1912-1918.

Minocci, A., G. Savia, R. Lucantoni, M.E. Berselli, M. Tagliaferri, G. Calo, M.L. Petroni, C. de Medici, G.C. Viberti and A. Liuzzi, 2000. Leptin plasma concentrations are dependent on body fat distribution in obese patients. International Journal of Obesity and Related Metabolic Disorders, Volume 24, 1139-1144.

Morrison, C.D., J.A. Daniel, B.J. Holmberg, J. Djiane, N. Raver, A. Gertler, D.H. Keisler, 2001. Central infussion of leptin into well-fed and undernourished ewe lambs: Effects on feed intake and serum concentrations of growth hormone and luteinizing hormone. Journal of Endocrinology, Volume 168, 317-324.

Oberkofler, H., A. Beer, D. Breban, E. Hell, F. Krempler and W. Patsch, 1997. Human obese gene expression: alternative splicing of mRNA and relation to adipose tissue localization. Obesity Surgery, Volume 7, 390-396.

Oldham, J.M., J.A. Martyn, S.P. Kirk, J.R. Napier and J.J. Bass, 1996. Regulation of type 1 insulin-like growth factor (IGF) receptors and IGF-I mRNA by age and nutrition in ovine skeletal muscles. Journal of Endocrinology, Volume 148, 337-346.

Peter, M.A., K.H. Winterhalter, M. Böni-Schnetzler, E.R. Froeschand J. Zapf, 1993. Regulation of insulin-like growth factor-I (IGF-I) and IGF binding proteins by growth hormone in rat white adipose tissue. Endocrinology, Volume 133, 2624-2631.

Ramachandran, A., C. Snehalatha, V. Vijay, K. Satyavani, E. Latha and S.M. Haffner, 1997. Plasma leptin in non-diabetic Asian Indians: association with abdominal adiposity. Diabetic Medicine, Volume 14, 937-941.

Ren, M.Q., J. Wegner, O. Bellmann, G.A. Brockmann, F. Schneider, F. Teuscher and K. Ender, 2002. Comparing mRNA levels of genes encoding leptin, leptin receotor, and lipoprotein lipase between dairy and beef cattle. Domestic Animal Endocrinology, Volume 23, 371-381.

Sauerwein, H., U. Heintges, M. Hennies, T. Selhorst and A. Daxenberger, 2004. Growth hormone induced alterations of leptin serum concentrations in dairy cows as measured by a novel enzyme immunoassay. Livestock Production Science, 87, 189-195.

Sjögren, K., J.L. Liu, K. Blad, S. Skrtic, O. Vidal, V. Wallenius, D. LeRoith, J. Törnell, O.G.P. Isaksson, J.O. Jansson and C. Ohlson, 1999. Liver-derived insulin-like growth factor I (IGF-I) is the principal source of IGF-I in blood but it is not required for postnatal body growth in mice. Proceeding of the National Academy of Science of the United States of America, Volume 96, 7088-7092.

Suttie, J.M., B.A. Veenvliet, R.P. Littlejohn, P.D. Gluckman, I.D. Corson and P.F. Fennessy, 1993. Growth hormone pulsatility in ram lambs of genotypes selected for fatness or leanness. Animal Production, Volume 57, 119-125.

Svensson, J., B. Söderpalm, K. Sjögren, J. Engel and C. Ohlsson, 2005. Liver-derived IGF-I regulates exploratory activity in old mice. American Journal of Physiology - Endocrinology and Metabolism, Volume 289, E466-E473.

Te Pas, M.F.W., C.L.M. Gerritsen, A.H. Visscher and K.H. De Greef, 2003. Relationships between performance traits and the expression of growth hormone, insulin-like growth factor-I, and insulin in pigs selected for growth or leanness. Journal of Animal Breeding and Genetics, Volume 120, 346-357.

Tiwary, C.M., 1986. An inexpensive method for assessing pituitary-response to luteinizing hormone releasing hormones - analysis of serum gonadotropins in pooled samples. Hormone and Metabolic Research, Volume 18, 616-620.

Tokuda, T., T. Matsui, J. Ito, S. Torii and H. Yano, 2000. The changes in body weight and plasma metabolite levels during leptin injection are caused by the reduction of food intake in sheep. Animal Science, Volume 71, 343-348.

Villafuerte, B.C., J.B. Fine, Y. Bai, W. Zhao, S. Fleming and M. DiGirolamo, 2000. Expressions of leptin and insulin-like growth factor-I are highly correlated and region-specific in adipose tissue of growing rats. Obesity Research, Volume 8, 646-655.

Wood, B.J., J.A. Archer and J.H.J. van der Werf, 2004. Response to selection in beef cattle using IGF-I as a selection criterion for residual feed intake under different Australian breeding objectives. Livestock Production Science, Volume 91, 69-81.

Wylie, A.R.G., D.M.B. Chestnutt and D.J. Kilpatrick, 1997. Growth and carcass characteristics of heavy slaughter weight lambs: effects of sire breed and sex of lamb and relationships to serum metabolites and IGF-I. Animal Science, Volume 64, 309-318.

Yamada, T., S.I. Kawakami and N.T. Nakanishi, 2003. The relationship between leptin concentrations and the distribution of body fat in crossbred steers. Animal Science Journal, Volume 74, 95-100.

Yang, S.H., T. Matsui, H. Kawachi, T. Yamada, N. Nakanishi and H. Yano, 2003. Fat depot-specific differences in leptin mRNA expression and its relation to adipocyte size in steers. Animal Science Journal, Volume 74, 17-21.

Relationship of live animal scores/measurements and carcass grades with carcass composition and carcass value of steers

Michael J. Drennan, M.G. Keane and M. McGee
Teagasc, Grange Beef Research Centre, Dunsany, Co. Meath, Ireland

Abstract

Steers (n = 134) slaughtered at 2 years of age were used. Pre-slaughter, ultrasonically scanned muscle depth at the 3rd lumbar vertebra and fat depth at the 3rd lumbar and 13th rib were recorded and visual muscular and skeletal scores were assigned. Carcass weights, conformation and fat scores were obtained post- slaughter. The right side of each carcass was dissected into meat, fat and bone. Carcass value (c/kg) was calculated as the sum of the commercial values of each boneless, fat trimmed meat cut. Muscular scores and scanned muscle depth showed positive correlations (r = 0.61 to 0.69) with carcass meat proportion, proportion of high-value cuts (HVC) and carcass value and negative correlations with carcass bone (r = ~-0.80) and fat (r = ~-0.33) proportions. The only significant correlations with scanned fat depth were carcass bone (r = -0.45) and fat (r = 0.30) proportions. Correlations with skeletal scores were negative (r = ~-0.4) for meat proportion, proportion of HVC and carcass value, and positive (r = ~0.3) for bone and fat proportions. Muscular score, muscle depth and fat depth combined explained ~0.62 of total variation in carcass meat proportion and value. Correlations between carcass conformation score and meat proportion, proportion of HVC and carcass value were positive (r = 0.73 to 0.82), and negative for bone (r = -0.82) and fat (r = -0.49) proportions. Corresponding correlations with carcass fat score were negative (r = ~-0.3) except for carcass fat proportion (r = 0.60). Carcass conformation and fat scores combined explained ~0.75 of total variation in carcass meat proportion and value. A unit (scale 1 to 15) increase in conformation increased carcass meat proportion and value by 14.1 g/kg and 6.4 c/kg, respectively. Corresponding values for a unit increase in fat score were -7.4 g/kg and -2.8 c/kg.

Keywords: beef, carcass composition, carcass grading, ultrasound measurements

Introduction

For beef cattle genetic improvement programmes, estimates of meat yield, meat distribution in the carcass (due to the differences in value between cuts) and ideally carcass value based on the latter are desirable. Live animal indicators of carcass meat yield and distribution include visual muscular and skeletal scoring, and ultrasonic scanning, for muscle size, subcutaneous fat depth and marbling. Carcass indicators are conformation and fat scores. In the European Union (EU) beef carcasses are graded for conformation (EUROP scale with E best for conformation), fatness (scale 1 to 5 with 5 fattest), sex category (steer, heifer, young bull, cow, bull) and carcass weight. Conformation and fatness are based on visual examination of carcasses (Commission of the European Communities 1982) which has now been replaced in Ireland by mechanical grading (Allen and Finnerty, 2000). Mechanical grading, in addition to ensuring greater consistency, facilitates grading on a 15-point rather than a 5-point scale. This is beneficial as carcasses in Ireland fall into a narrow range of grades. For example, conformation grades R and O accounted for 87% of steers and 91% of heifers in 2005 (Drennan, 2006). Likewise, fat grades 3 and 4 (4L, 4 and 4H) accounted for 88 and 85% of steer and heifer carcasses, respectively.

Tatum *et al.* (1986) showed that a subjective score for muscle thickness on the live animal was significantly associated with meat yield. Perry *et al.* (1993a) showed that accurate visual scoring on the live animal can be achieved and reported a correlation of 0.70 between saleable meat and muscle score.

Greiner *et al.* (2003) concluded from their studies that prediction equations derived from live animal ultrasound measurements are similar in accuracy to carcass measurements for predicting carcass composition. In fact both Bergen *et al.* (2005) and Tait *et al.* (2005) found that equations based on live animal measurements gave more precise predictions of lean meat yield than equations derived from carcass measurements. Herring *et al.* (1994) concluded that ultrasound *longissimus* muscle area may be a more accurate estimate of actual *longissimus* muscle than that obtained by carcass measurement which may be distorted by angle of ribbing and different body pressures. In studies (Hamlin *et al.*, 1995; Greiner *et al.*, 2003) using live animal ultrasound scanning fat thickness explained about 0.6 of the total variation in carcass meat percentage while the contribution of *longissimus* muscle area was minimal.

Earlier studies using scores for carcass conformation (shape due to muscle and fat layers) were shown to have little value as an indictor of meat yield but that fat-corrected conformation score is of value in commercial classification schemes (Kempster and Harrington, 1980; Kempster, 1986). Studies by Perry *et al.* (1993b) showed that carcass weight alone, carcass weight with carcass muscle score, and carcass weight with carcass muscle and fat scores, accounted for 0.1, 37.9 and 46.7% respectively of the total variation in saleable meat yield.

The objective of the present study was to examine the relationships of live animal visual muscular and skeletal scores, ultrasound muscle and fat measurements, and carcass conformation and fat scores, with carcass composition and estimated carcass value of steers.

Materials and methods

A total of 134 steers, slaughtered at approximately two years of age, were used in the study. They comprised of 65 Holstein-Friesians (HF), 16 Limousin X HF, 16 Belgian Blue X HF, 17 Aberdeen Angus X HF and 20 $\geq$ 0.75 late maturing breed crosses. They were slaughtered in three separate batches of 44, 47 and 43 animals each.

Prior to slaughter, visual scores were assigned to each animal using both the Irish Cattle Breeding Federation (ICBF) and Signet (Allen, 1990) scoring procedures. The ICBF system involved assigning muscular scores (scale 1 to 15) at six locations (width at withers, width behind withers, loin development, development of hindquarters, width of hindquarters and development of the inner thigh) and skeletal scores (scale 1 to 10) at three locations (length of back, pelvic length and height at withers). Two ICBF assessors were used to score the animals on each occasion, with assessor A scoring all the animals on all occasions and assessors B or C scoring on different occasions. In the Signet procedure, muscular scores (scale 1 to 15) were assigned at 3 locations (roundness of hindquarters, width of rump and width and thickness of the loin) using two assessors (X and Y) from the research centre. For each assessor the scores at the different locations recorded were averaged to give one score for each animal. Correlations for muscular scores between the ICBF assessors and assessor X using the Signet procedure varied from 0.91 to 0.93 while the correlations between assessor Y and the others varied from 0.73 to 0.80. In addition, the animals were ultrasonically scanned for eye muscle depth at the 3rd lumbar vertebra and for fat depth at both the 3rd lumbar vertebra (3 sites) and, at the 13th rib (4 sites) using a Dynamic Imaging scanner (Concept/MCV Veterinary Ultrasound Scanner).

Hot carcass weight was recorded at slaughter, and perinepheric plus retroperitoneal fat was weighed. Cold carcass weight was taken as 0.98 hot carcass weight. Carcasses were mechanically (Allen and

Finnerty, 2000) and visually graded according to the EU Beef Carcasses Classification Scheme (Commission of the European Communities, 1982). Carcass meat, fat and bone proportions were obtained from dissection of the right side of each carcass following quartering into an 8 rib pistola hindquarter and the remaining forequarter. The pistola was dissected into 12 cuts (silverside, topside, knuckle, rump, tail of rump, fillet, striploin, cube roll, cap of ribs, leg, heel and eye of round). The bones were removed and scraped clean. All visible dissectable fat was removed from each cut. The weight of each cut and total weight of fat trim, lean trim and bone were recorded. The forequarter was dissected into 11 cuts (front shin, brisket, chuck, neck, flat ribs (1 to 5), plate, *triceps brachii*, bladesteak, braising muscle, chuck tender and clod) and a similar procedure was undertaken as for the pistola. For both quarters, lean trim was added to the fat trimmed boneless cuts to give meat yield. Total half carcass yields of meat, fat and bone were the combined values for the pistola and forequarter. Carcass value was taken as the sum of the commercial values of the boneless, fat trimmed cuts from the half carcass with a small deduction for bone expressed as a proportion of half carcass weight. Fat was given zero value when estimating carcass value. Data was analysed using PROC CORR and PROC REG procedures of the Statistical Analysis System Institute (SAS, 2003).

Results

Experimental animals

Details of the animals are presented in Table 1. Average liveweight at slaughter was 619 kg and carcass weight was 319 kg. Carcass conformation and fat scores (scale 1 to 15) obtained using mechanical grading were 5.7 and 8.7, respectively. Carcass meat, fat and bone proportions were 672, 120 and 208 g per kg, respectively. The proportion of high value cuts (fillet, striploin and cube roll) in the carcass was 70 g/kg.

Correlations of live animal scores/measurements with carcass composition, value and grades

The correlations of the visual muscular scores, using the ICBF scoring procedure pre-slaughter, with carcass meat, fat and bone proportions, the proportion of high-value cuts and carcass value was similar between assessors A and B/C. Thus, only the results for assessor A who scored all animals are shown (Table 2). There were high positive correlations ($P<0.001$) between average (6 locations) muscular score and carcass meat proportion ($r = 0.69$), proportion of high-value cuts ($r = 0.61$), carcass value ($r = 0.69$) and carcass conformation score ($r = 0.87$). High negative correlations ($P<0.001$) were obtained between the muscular score and carcass bone proportion ($r = -0.83$). Correlations with carcass fat proportion were low and negative ($r = -0.34$). Similar results were obtained for the ICBF and Signet scoring systems (Table 2). Correlations between each of the three individual ICBF skeletal scores and meat proportion, proportion of high-value cuts, carcass value and carcass conformation were low and negative ($r = -0.26$ to -0.51). Correlations between the skeletal scores and carcass bone proportion, although significant were low and positive ($r = 0.36$ to 0.45), while those with carcass fat proportion and carcass fat score were also positive and low ($r = 0.11$ to 0.29). Correlation with the average of all three skeletal scores were generally lower than those with individual scores. High correlations ($P<0.001$) were obtained between scanned muscle depth and carcass meat proportion ($r = 0.66$), proportion of high-value cuts ($r = 0.63$), carcass value ($r = 0.69$) and carcass conformation score ($r = 0.79$), while a negative ($P<0.001$) correlation ($r = -0.79$) was obtained with carcass bone proportion. When scanned muscle depth was expressed relative to carcass weight all correlations were substantially reduced. Correlations of carcass weight with the various carcass traits were moderate. Correlations of scanned fat depth with meat proportion, proportion of high-value cuts and carcass value were not significant ($r = 0.03$ to 0.07). The correlations of fat depth with carcass fat proportion and carcass fat score were 0.30 and 0.54, respectively.

Table 1. Mean, standard and deviation, range and coefficient of variation for live animal and carcass data for 134 steers.

		Mean	Std. dev.	Minimum	Maximum	Coefficient of variations
Pre-slaughter						
Scanned muscle depth (mm)		63	8.7	45	89	14.0
Scanned fat depth (mm)		1.3	0.813	0.33	4.4	62.1
Muscular score A[a]		6.8	2.11	1.8	10.7	30.8
Muscular score B/C[a]		6.5	2.44	2.3	13.2	37.3
Muscular score X[b]		4.6	2.20	1.0	10.3	48.3
Muscular score Y[b]		4.7	2.76	1.0	11.0	58.9
Slaughter wt. (kg)		619	62.1	495	784	10.0
Post-slaughter						
Cold carcass weight (kg)		319	36.3	257	450	11.3
Conformation score[c]	-mechanical	5.7	2.17	2.0	11.0	38.1
	-visual	6.0	2.47	2.0	11.0	41.1
Fat score[d]	-mechanical	8.7	1.81	4.0	14.0	20.8
	-visual	9.2	1.62	5.0	14.0	17.4
Meat (g/kg)		672	38.2	593	771	5.7
Fat (g/kg)		120	27.7	55	211	23.0
Bone (g/kg)		208	20.6	160	260	9.9
High-value cuts[e] (g/kg)		70	5.7	56	87	81
Value (c/kg)		247	17	210	292	681

[a]Assessor A and B/C using the ICBF procedure (scale 1 to 15).
[b]Assessors X and Y using the Signet muscular scoring procedure (scale 1 to 15).
[c]Scale 1 to 15 (best conformation)
[d]Scale 1 to 15 (fattest).
[e]Meat in fillet, striploin and cube roll.

Regression equations relating live animal data to carcass composition, value and grades

Regression equations describing the relationship between scanned muscle and fat measurements and muscular scores using the ICBF procedure with carcass meat, fat and bone proportions, meat distribution and estimated carcass value are shown in Table 3. The combined muscular scores and scanned muscle and fat depth measurements explained a high proportion of the total variation in percent meat yield, percent high-value cuts in the carcass and carcass value ($R^2 = 0.49$ to 0.63). The corresponding R^2 values with carcass bone and fat proportions were 0.76 and 0.37, respectively (Table 3). Using the combined scanned measurements and muscular scores resulted in higher R^2 values than those obtained with either alone.

Correlations of carcass grades with carcass composition and value

Using mechanical grading, high positive correlations ($r = 0.73$ to 0.82) were obtained for carcass conformation score with carcass meat proportion, the proportion of high-value meat cuts and carcass value (Table 4). A high negative correlation ($r = -0.82$) was obtained between carcass conformation and carcass bone proportion. The correlation between carcass conformation score and carcass fat proportion was low and negative. The only positive correlation with carcass fat score was the proportion of carcass fat ($r = 0.60$) and high-value cuts as a proportion of total meat ($r = 0.34$).

Table 2. Correlations between live animal visual muscular and skeletal scores and scanned muscle and fat measurements with carcass composition, value and grades.

	Carcass			High-value cuts	Carcass value	Carcass grades[c]	
	Meat	Fat	Bone			Conf.	Fat
Muscular scores[a]							
Av: 6 locations	0.69***	-0.34***	-0.83***	0.61***	0.69***	0.87***	-0.00
Av: 2 locations[b]	0.69***	-0.37***	-0.78***	0.58***	0.68***	0.86***	-0.07
Skeletal scores[a]							
Height at withers	-0.42***	0.29***	0.38***	-0.37***	-0.41***	-0.51***	0.27**
Length of back	-0.34***	0.21*	0.36***	-0.26**	-0.33***	-0.39***	0.11
Length of pelvis	-0.42***	0.24**	0.45***	-0.36***	-0.39***	-0.49***	0.12
Av. score	-0.33***	0.23**	0.29***	-0.29***	-0.30***	-0.35***	0.33***
Signet Assessor X	0.69***	-0.34***	-0.83***	0.61***	0.69***	0.87***	-0.00
Signet Assessor Y	0.66***	-0.30***	-0.81***	0.58***	0.69***	0.81***	0.19*
Scan. fat depth	0.03	0.30***	-0.45***	0.07	0.07	0.35***	0.54***
Scan. muscle depth	0.66***	-0.32***	-0.79***	0.63***	0.69***	0.79***	0.13
Carcass weight	0.39***	-0.09	-0.61***	0.37***	0.44***	0.55***	0.38***
Muscle depth/carc. wt.	0.42***	-0.30***	-0.37***	0.41***	0.41***	0.44***	-0.21*

[a]ICBF Assessor A.

[b]Width and roundness of hindquarters.

[c]Mechanical grading.

Table 3. Regression equations relating live animal muscular scoresa and ultrasound muscle and fat measurements to the proportions (g/kg) of meat, fat, bone and high-value meat cuts in the carcass and carcass value(c/kg).

	Intercept	Muscular score[b]	Muscle depth[c]	Fat depth[c]	R^2
Meat	586	+12.6 (1.14)***			0.48
	472		+3.5 (0.29)***	-15.3 (3.16)***	0.52
	509	+8.6 (1.47)***	+2.0 (0.37)***	-16.1 (2.83)***	0.62
Fat	151	-4.5 (1.08)***			0.11
	208		-1.8 (0.25)***	+18.7 (2.68)***	0.35
	194	-3.3 (1.37)*	-1.2 (0.34)***	+19.0 (2.64)***	0.37
Bone	263	-8.2 (0.47)***			0.69
	319		-1.7 (0.14)***	-3.34 (1.50)*	0.63
	297	-5.3 (0.63)***	-0.7 (0.158)***	-2.9 (1.21)*	0.76
High-value cuts	59	+1.6 (0.19)***			0.37
	42		+0.48 (0.047)***	-1.81 (0.506)***	0.45
	46	+0.88 (0.253)***	+0.33 (0.064)***	-1.89 (0.487)***	0.49
Value	209	+6.0 (1.0)			0.48
	155		+1.6 (0.13)***	-6.2 (1.4)***	0.55
	169	+3.3 (0.64)***	+1.0 (0.16)***	-6.5 (1.24)***	0.63

[a]ICBF Assessor A.

[b]Scale 1 to 15; [c]mm.

Table 4. Correlations between carcass conformation and fat score with carcass meat, fat and bone proportion, proportion of high-value meat cuts in the carcass, high-value meat cuts as a proportion of carcass meat and carcass value.

Grading method	Meat	Fat	Bone	High-value meat cuts	High-value cuts as % meat	Carcass value
Mechanical						
Conformation	0.80***	-0.49***	-0.82***	0.73***	0.59***	0.82***
Fat	-0.33***	0.60***	-0.20	-0.22*	0.34***	-0.28***
Visual						
Conformation	0.76***	-0.46***	-0.80***	0.68***	0.54***	0.79***
Fat	-0.32***	0.60***	-0.19*	-0.25**	0.21*	-0.31***

Visual and mechanical conformation scores were highly correlated (r = 0.91) but the corresponding correlation for fat score was lower (0.79). The relationship of various carcass traits, and particularly carcass value, was considerably better for conformation than for fat score.

Regression equations relating carcass grades to carcass composition and value

Regression equations describing the relationship between carcass conformation and fat scores with carcass meat, fat and bone proportions, meat distribution and carcass value are shown in Table 5. Using mechanical carcass grading, the effects of a one unit increase (scale 1 to 15) in carcass conformation and fat scores on carcass meat proportion were +14 and -7 g per kg, respectively. The corresponding effects on carcass fat proportion were -6 and +10 g/kg, on carcass bone proportions were -8 and -2 g/kg, and on high-value meat cuts in the carcass were +1.9 and -0.7 g/kg. The effects of conformation and fat scores on meat in high-value cuts as a proportion of total carcass meat were both positive. The effects on carcass value of a one unit increase in carcass conformation and fat scores were +6.4 and -2.8 c/kg, carcass, respectively. The R^2 values obtained with visual grading of the carcasses for conformation and fatness were somewhat lower than those obtained with mechanical grading.

Table 5. Regression equations using carcass conformation and fat scores (s.e.) for predicting carcass meat, fat and bone proportions (g/kg), the proportions of high-value meat cuts in the carcass, high-value cuts as a proportion of total carcass meat (g/kg) and carcass value (c/kg).

	Intercept	Conformation score	Fat score	R^2
Mechanical grading				
Meat	656	+14.1 (0.78)***	-7.4 (0.98)***	0.75
Fat	71	-6.2 (0.71)***	+9.8 (0.89)***	0.60
Bone	272	-7.8 (0.44)***	-2.4 (0.56)***	0.71
High-value cuts in carcass	65	+1.9 (0.15)***	-0.72 (0.19)***	0.57
High-value cuts in meat	93	+6.1 (0.65)***	+4.4(0.82)***	0.47
Value	235	+6.4 (0.34)***	-2.8 (0.43)***	0.76
Visual grading				
Meat	669	+11.7 (0.77)***	-7.2 (1.17)***	0.68
Fat	58	-4.9 (0.66)***	+10.0 (1.00)***	0.55
Bone	273	-6.7 (0.41)***	-2.7 (0.62)***	0.69
High-value cuts in carcass	68	+1.5 (0.14)***	-0.8 (0.21)***	0.51
High-value cuts in meat	107	+4.9 (0.63)**	3.0 (0.96)**	0.35
Value	242	+5.3 (0.33)***	-3.0 (0.50)***	0.70

Discussion

Assessors

The correlations for muscular score between both of the ICBF assessors and assessor X using the Signet procedure varied from 0.91 to 0.93. The corresponding correlations between these assessors and assessor Y varied from 0.73 to 0.80. This shows that a high level of consistency can be achieved where experienced assessors carry out these subjective evaluations. In similar evaluations Perry *et al.* (1993b) obtained correlations between three assessors ranging from 0.91 to 0.95.

Visual muscular scores

The very high correlation (r = 0.81 to 0.87) between muscular score at slaughter and carcass conformation score indicates that both evaluations give a similar description of the animal or carcass in terms of muscular development. Perry *et al.* (1993a, b) reported correlations of 0.84 and 0.79, respectively between live animal and carcass muscle scores. The fact that there were significant negative relationships (r = -0.30 to -0.37) between muscular scores pre-slaughter and carcass fat proportion indicates that assessors did not give steers with more fat a better muscular score. In the present study, high positive correlations were obtained for the muscular scores taken on the live animal with meat proportion, the proportion of high-value cuts in the carcass and carcass value. Negative correlations were obtained between the muscular scores and carcass bone and fat proportions. The most important evaluations involve relationships with saleable meat proportion and carcass value. The latter takes account of both saleable meat proportion and its distribution in the carcass. Perry *et al.* (1993a) reported a correlation of 0.70 between live animal muscle score and percent saleable meat, which is similar the findings in the present study, while May *et al.* (2000) obtained a considerably lower correlation (r = 0.35) with percentage yield of boneless subprimals. Tatum *et al.* (1986) in a study using 324 steers, found that within a frame size, the percent separable muscle in the carcass increased and percent bone decreased with increased muscle thickness when adjusted to a constant fat percentage.

Correlations of muscular scores with carcass meat proportion and value were equally as good using only two locations (width and roundness of the hindquarter) as those using the entire six locations (r = ~ 0.69). Similarly, the Signet procedure using only three locations (width and roundness of hindquarters and loin development) for muscular scoring also resulted in r values of approximately 0.69.

Visual skeletal scores

The main effects of skeletal scores were significant negative correlations with carcass meat proportion, proportion of high-value cuts in the carcass, carcass value and carcass conformation and significant positive correlations with carcass bone and fat proportions. In contrast, May *et al.* (2000) found that percent boneless subprimals increased with increasing frame size (r = 0.26). Bergen *et al.* (2005) found that hip height was positively associated with lean meat yield, explaining 5 to 6% of the variation in yield but heart girth, body length and hindquarter width did not contribute to predictions of carcass composition.

Ultrasonic scanning for muscle and fat

A Dynamic Imaging scanner was used in the present study to measure muscle depth at the 3rd lumbar vertebra whereas in most published studies eye muscle area was recorded at the 12th/13th rib using an Aloka ultrasound unit. Bergen *et al.* (2003) found that *longissimus* muscle depth and width measurements were as valuable as *longissimus* area for predicting carcass composition of yearling beef bulls while Bergen *et al.* (2005) reported that meat yield can be accurately predicted using fat and *longissimus* muscle depth measurements.

Correlations of 0.66 and 0.69 were recorded between scanned muscle depth with carcass meat proportion and value, respectively but similar correlations with scanned fat depth gave values close to zero. Perry *et al.* (1993a) obtained correlations of 0.58 and -0.59 between eye muscle area at the $12^{th}/13^{th}$ rib recorded on the carcass and scanned rump (P8) fat depth recorded on the live animal, respectively with percent saleable meat. Griffin *et al.* (1999) and May *et al.* (2000) obtained low correlations (r = 0.07) between ultrasound *longissimus* muscle area and percentage yield of boneless subprimals. In most studies, significant negative correlations were obtained between ultrasound fat depth measurements and percent carcass meat with Griffin *et al.* (1999) and May *et al.* (2000) reporting values of -0.40 and -0.73, respectively.

Carcass grades

Correlations of carcass conformation using mechanical grading and carcass meat proportion and value were 0.80 and 0.82, respectively. Perry *et al.* (1993a) obtained a correlation of 0.60 between carcass muscle score and saleable meat while Perry *et al.* (1993b) found that at the same carcass weight and fat depth there was an increase of 1.9% in saleable meat per unit improvement (scale 1 to 5) in carcass muscle score.

Regression of carcass composition and value on live animal measurements

In the present study, the proportion of total variation in carcass meat proportion explained by muscular score (assessor A using the ICBF procedure $R^2 = 0.48$) was similar to that obtained using scanned muscle and fat depth measurements ($R^2 = 0.52$). Combining the muscular score and scanned measurements increased the proportion of total variation accounted for in meat yield to 0.62. Perry *et al.* (1993a) found that liveweight, liveweight with live muscle score and liveweight with muscle score and fat depth accounted for 0.4, 49.7 and 61.6% respectively of the total variation in percent saleable meat yield. Corresponding values reported by Perry *et al.* (1993b) were 0.1, 37.4 and 46.3. May *et al.* (2000) found that frame size, muscle score, fat thickness and liveweight explained 66% of the variation in the percentage yield of boneless subprimals. Bergen *et al.* (1996) found that ultrasound muscle and fat measurements explained 0.73 of the variation in carcass lean yield. The latter authors stated that as the sole predictor of lean content, scanned fat had an R^2 of 0.59 and the smaller regression coefficient for *longissimus dorsi* area means that it was less important than fat and thus, should not be used as the sole predictor of lean content. Tait *et al.* (2005) reported that ultrasound fat thickness at the $12^{th}/13^{th}$ rib explained 0.32 of total variation in percentage retail product which was increased by a further 0.09 with the inclusion of ultrasound *longissimus* muscle area at the same location. Bergen *et al.* (2005) found that of the 0.55 of lean meat yield predicted from live animal measurements, rib fat depth, *longissimus* muscle depth, hip height and marbling score accounted for 0.32, 0.14, 0.05 and 0.04, respectively.

Hamlin *et al.* (1995) reported that ultrasonically scanned fat thickness alone explained up to 0.59 of the total variation in retail meat yield percentage which agrees with other findings ($R^2 = 0.50$, Epley *et al.,* 1970; $R^2 = 0.51$, Wallace *et al.,* 1977; $R^2 = 0.54$, Greiner *et al.,* 2003). All of these studies reported that ultrasonically measured *longissimus* muscle area was not an accurate predictor of retail yield percentage or at best only accounted for a small proportion of total variation in yield. In contrast, in the present study, when used in regression equations for carcass meat proportion and value, both muscular score and scanned muscle depth had significant positive effects while scanned fat depth had significant negative affects. The greater importance of scanned muscle depth in predicting carcass meat yield and carcass value in the present study is probably due to animals with better muscular scores having heavier carcasses and therefore greater scanned muscle depths than those with poorer muscular scores. At one extreme the steers of ≥ 0.75 late maturing breed types had carcass weights, muscle depths and meat proportions of 376 kg 77 mm and 723 g/kg, respectively. The corresponding figures for HF at the other extreme were 310 kg, 57 mm and 648 g/kg.

Regression of carcass composition, value and other traits on carcass grades

As beef carcasses in the EU are graded for conformation and fatness as part of the carcass classification system it would be desirable to be able to predict important traits from carcass grades. The two most important traits are carcass meat yield and carcass value. In the present study, using mechanical grading for conformation and fatness accounted for 0.75 and 0.76 of the total variation in meat yield and value, respectively. Bohuslávek (2002) obtained correlations between carcass conformation and fat scores with percentage saleable meat of 0.05 and -0.71, respectively. However, the author reported that when examined within fat classes, significant correlations between conformation and percentage meat yield were obtained with a value of 0.595 recorded with fat class 3 animals. Perry *et al.* (1993a) found that carcass weight alone, carcass weight with carcass muscle score and carcass weight with carcass muscle score and fat depth accounted for 3.7, 37.6 and 52.2%, respectively of the variation in percent saleable meat yield. Corresponding values reported by Perry *et al.* (1993b) were 0.1, 37.9 and 46.7%.

Jones *et al.* (1989) found that a visual overall fat score, a muscle thickness score and the combined fat and muscle scores taken on the cold carcass in a model that included breed, sex and interactions explained 0.42, 0.20 and 0.51 respectively of total variation in lean meat yield. Herring *et al.* (1994) found that carcass fat thickness and kidney, heart and pelvic heart were the two most important predictors of carcass cutability.

In the present study, a 1 unit increase in carcass conformation score (scale 1 to 15) was associated with an increase in carcass meat proportion and value by 14.1 g/kg and 6.4 c/kg respectively. The corresponding effects of a 1 unit (scale 1 to 15) increase in carcass fat score were -7.4 g/kg and -2.8 c/kg. Values for R^2 were marginally lower using visual grading compared with those obtained using mechanical grading.

Although the regression equations presented give an indication of expected outcomes of meat yield and carcass value using both the live animal scores/measurements and carcass grades the sample of animals needs to be enlarged to be representative of the entire population before these could be used as prediction equations.

Conclusion

The results show that both live animal evaluations and carcass grades can be effective in predicting carcass meat proportion and carcass value. Carcass grades are however more effective as predictors explaining about 0.75 of total variation in meat yield and value compared to approximately 0.62 for live animal evaluations. In live animal and carcass regression equations the effects of both muscular or conformation scores and fat measurements were highly significant in estimating carcass meat, fat and bone proportions, the proportion of high-value cuts and carcass value.

Acknowledgements

Thanks are due to Mr. B. Davis, Mr. J. Marron and Mr. M. Fagan for valuable technical assistance and Mrs. M. Weldon for typing the document.

References

Allen, D., 1990. Planned beef production and marketing. Blackwell Science Publ., London, UK, 248p, 143.

Allen, P. and N. Finnerty, 2000. Objective beef carcass classification. A report of a trial of three VIA classification systems. Teagasc and Department of Agriculture, Food and Rural Development publication, 34 p.

Bergen, R.D., J.J. McKinnon, D. A. Christensen and N. Kohle, 1996. Prediction of lean yield in yearling bulls using real-time ultrasound. Canadian Journal of Animal Science, Volume 76, 305-311.

Bergen, R., D.H. Crews, S.P. Miller and J.J. McKinnon, 2003. Predicting lean meat yield in beef cattle using ultrasonic muscle depth and width measurements. Canadian Journal of Animal Science, Volume 83, 429-434.

Bergen, R., S.P. Miller, I.B. Mandell and W.M. Robertson, 2005. Use of live ultrasound, weight and linear measurements to predict carcass composition of young beef bulls. Canadian Journal of Animal Science, Volume 85, 23-35.

Bohuslávek, Z., 2002. Analysis of the commercial grades for beef carcases. Czech Journal of Animal Science, Volume 47, 112-118.

Commission of the European Communities, 1982. Commission of the European Communities (Beef Carcass Classification) Regulations. Council Regulations 1358/80, 1208/81, 1202/82. Commission Regulations 2930/81, 563/82, 1557/82, Commission of the European Communities, Brussels, Belgium.

Drennan, M.J., 2006. Relationship between beef carcass classification grades with meat yield and value. Irish Grassland Association Journal, Volume 40, (in press).

Epley, R.J, H.B. Hedrick, W.C. Stringer and D. P. Hutcheson, 1970. Prediction of weight and percent retail cuts of beef using five carcass measurements. Journal of Animal Science, Volume 30, 872-879.

Greiner, S.P., G.H. Rouse, D.E. Wilson, L.V. Cundiff and T.L. Wheeler, 2003. Prediction of retail product weight and percentage using ultrasound and carcass measurements in beef cattle. Journal of Animal Science, Volume 81, 1736-1742.

Griffin, D.B., J.W. Savell, H.A. Recio, R.P. Garrett and H.R. Cross, 1999. Predicting carcass composition of beef cattle using ultrasound technology. Journal of Animal Science, Volume 77, 889-892.

Hamlin, K.E., R.D. Green, L.V. Cundiff, T.L. Wheeler and M.E. Dikeman, 1995. Real-time ultrasonic measurement of fat thickness and *longissimus* muscle area: II. Relationship between real-time ultrasound measures and carcass retail yield. Journal of Animal Science, Volume 73, 1725-1734.

Herring, W.O., S.E. Williams, J.K. Bertrand, L.L. Benyshek and D.C. Miller, 1994. Comparison of live and carcass equations predicting percentage of cutability, retail product weight and trimmable fat in beef cattle. Journal of Animal Science, Volume 72, 1107-1118.

Jones, S.D.M., A.K.W. Tong and W.M. Robertson, 1989. The prediction of beef carcass lean content by an electronic probe, a visual scoring system and carcass measurements. Canadian Journal of Animal Science, Volume 69, 641-648.

Kempster, A.J., 1986. Estimation of the carcass composition of different cattle breeds and crosses from conformation assessments adjusted for fatness. Journal of Agricultural Science Cambridge, Volume 106, 239-254.

Kempster, A.J. and G. Harrington, 1980. The value of 'fat corrected' conformation as an indicator of beef carcass composition within and between breeds. Livestock Production Science, Volume 7, 361-372.

May, S.G., W.L. Mies, J.W. Edwards, J.J. Harris, J.B. Morgan, R.P. Garrett, F.L. Williams, J. W. Wise, H.R. Cross and J. W. Savell, 2000. Using live estimates and ultrasound measurements to predict beef carcass cutability. Journal of Animal Science, Volume 78, 1255-1261.

Perry, D., W.A. McKiernan and A.P. Yeates, 1993a. Muscle score: its usefulness in describing the potential yield of saleable meat from live steers and their carcasses. Australian Journal of Experimental Agriculture, Volume 33, 275-281.

Perry, D., A.P. Yeates and W.A. McKiernan, 1993b. Meat yield and subjective muscle scores in medium weight steers. Australian Journal of Experimental Agriculture, Volume 33, 825-831.

Statistical Analysis Systems Institute, 2003. SAS/STAT. SAS systems for windows. Release 9.1.3, SAS Institute Inc., Cary, NC, USA.

Tait, R.G. Jr, D.E. Wilson and G.H. Rouse, 2005. Prediction of retail product and trimmable fat yields from the four primal cuts in beef cattle using ultrasound or carcass data. Journal of Animal Science, Volume 83, 1353-1360.

Tatum, J.D., H.G. Dolezal, F.L. Williams Jr, R. A. Bowling and R.E. Taylor, 1986. Effects of feeder-cattle frame size and muscle thickness on subsequent growth and carcass development. II. Absolute growth and associated changes in carcass composition. Journal of Animal Science, Volume 62, 121-131.
Wallace, M.A., J.R. Stouffer and R.G. Westervelt, 1977. Relationships of ultrasonic and carcass measurements with retail yield in beef cattle. Livestock Production Science, Volume 4, 153-164.

Use of carcass weight, community scale for carcass classification and carcass ultrasound measurements to predict carcass composition of young beef bulls

Rafael Delfa, G. Ripoll, B. Panea, M. Joy and P. Albertí
Unidad de Tecnología en Producción Animal, CITA de Aragon, P.O. Box 727, 50080 Zaragoza, Spain

Abstract

Sixty-nine carcasses of young bulls of four different Spanish local breeds with a hot carcass weight ranging from 199.9 to 449.6 Kg were classified by the Community Scale Carcass Classification and scanned by ultrasound to determine *M. longissimus thoracis* depth (BCAR) and subcutaneous fat thickness (CCAR). Carcasses were halved and cooled at 4 °C for 24 h. The left side was divided in standardised commercial joints and each joint was cut into retail muscle, subcutaneous and inter-muscular trim fat, and bone. The hot carcass weight, carcass classification and ultrasound measurements were used to predict carcass tissue composition by Stepwise regression analysis. The developed model with hot carcass weight and carcass classification explained 97, 60 and 85% of the muscle, fat and bone weight variation, respectively. When ultrasound measurements were included into the above model, it was observed an increase of percentage of variation explained for 1 and 23 percent units for muscle and fat weight, respectively. For bone weight, the RSD was reduced by 8.8%. Results indicate that the used methodology could be an acceptable predictor of carcass tissue composition.

Keywords: ultrasound, beef, prediction, muscle depth, fat thickness

Introduction

The animal's economic value depends on carcass tissue composition, sealable meat, carcass fat distribution, and being these parameters the main criteria in carcass quality evaluation (Delfa, 1992; Delfa *et al.*, 1992). Therefore, it is necessary to develop an accurate method of carcass classification that allows assessing objectively carcass composition.

Nowadays, one of the most promising techniques to assess carcass composition is Real Time Ultrasound Scanner because it presents a high potential for the immediate use (Delfa *et al.*, 2005; Teixeira *et al.,* 2006). This tool can predict with high precision the tissue and anatomical composition of carcasses, which are carried out *in vivo or in hot carcass*. Wilson (1992) stated that considerable research and development was needed before effective use of ultrasound could be applied in cattle production and breeding. Since then, ultrasound technology has been developed and described in numerous publications (Wallace *et al.*, 1997; Larsgard and Kolstard, 2003; Greiner *et al.*, 2003; Teixeira *et al.*, 2006).

The main objective of the present study was to determine the precision of carcass weight, SEUROP grading system and ultrasound measures as predictors of carcass composition.

Material and methods

Sixty-nine carcasses of young bulls of four different Spanish local breeds (29 Asturiana de los Valles, 31 Asturiana de las Montañas, 23 Avileña- Negra Ibérica y 28 Pirenaica) were fed *ad libitum* with concentrate and cereal straw until they reached to 15 months of age. Animals were slaughtered at nearest commercial abattoir. Immediately after slaughter hot carcass weight (HWC) was recorded. An ALOKA SSD-900 Real Time Ultrasound Scanner (Toshiba Sonolayer scanner, model Sal-32B, with a 7.5 MHz probe) was used to determine *M.* l*ongissimus thoracis* depth (BCAR) and subcutaneous fat thickness (CCAR) at 13th thoracic vertebrae anatomical position.

Afterwards, carcasses were halved, chilled at 4ºC for 24 h and graded by the Community Scale Carcass Classification. The conformation (CONF) was scored, with grade values oscillating between 18 for S+ (excellent) to 1 for P- (poor), and the classification for fatness degree between 15 for 5+ (very high) to 1 for 1- (very low). Finally, the left side of the carcass was divided in standardised commercial joints and each joint was cut into retail muscle, subcutaneous and inter-muscular trim fat, and bone.

Data were analysed by stepwise regression (Ott, 1993) using the ultrasonic measurement, hot carcass weight and carcass classification as independent variables. The statistics R^2 and residual standard deviation (RSD) were used to assess the accuracy of prediction of ultrasonic measurement and hot carcass weight (Kempster *et al.*, 1982).

Results and discussion

Hot carcass weight ranged from 199.9 kg to 499.6 kg and SEUROP classification ranged from 4 (O^-) to 17 (S) for conformation and from 2 (1) to 8 (3) for fat cover, with means of 9.8 (U^-) and 5.1 (2), being a representative population of Spanish beef market.

The percentages of variation (R^2) for carcass tissue composition accounted for by carcass weight and SEUROP classification are given in Table 1. In Table 2 are shown the same percentages of variation (R^2) when ultrasonic measurements were included in the model. All developed models were highly significant ($p<0.001$).

Table 1. Determination coefficients (R2), residual standard deviation (RSD) and prediction equations for carcass tissue composition (muscle, fat and bone) from hot carcass weight (HCW) and SEUROP classification.

Step	Dependent variable (y)	Independent variable (x)	R^2	RSD	b	Sb	Intercept
1	Muscle weight	HCW	0.93***	11.930	0.56	0.015	-15.65
2		CONF	0.97***	7.642	3.83	0.127	
3		FATNESS	0.97**	7.172	-2.13	0.677	
1	Log 10 Fat weight	FATNESS	0.32***	0.140	0.04	0.010	0.85
2		HCW	0.39***	0.133	0.03	0.007	
3		CONF	0.60***	0.108	-0.05	0.006	
1	Bone weight	HCW	0.76***	2.281	0.09	0.004	4.66
2		CONF	0.85***	1.802	-0.59	0.106	
3		FATNESS	0.85*	1.787	0.28	0.169	

*$p\leq0.05$; **$p\leq0.01$; ***$p\leq0.001$.

Table 2. Determination coefficients (R2), residual standard deviation (RSD) and prediction equations for carcass tissue composition (muscle, fat and bone) from hot carcass weight (HCW) SEUROP classification and carcass ultrasound measurements.

Step	Dependent variable (y)	Independent variable (x)	R^2	RSD	b	Sb	Intercept
1	Muscle weight	HCW	0.94***	11.757	0.60	0.017	-14.34
2		CONF	0.98***	7.214	2.86	0.441	
3		CCAR	0.98***	6.316	-0.16	0.033	
4		FATNESS	0.98**	6.002	-1.88	0.664	
1	Log 10 Fat weight	CCAR	0.53***	0.128	0.01	0.001	0.67
2		FATNESS	0.67***	0.107	0.04	0.009	
3		HCW	0.72***	0.099	0.01	0.001	
4		CONF	0.83***	0.077	-0.04	0.006	
1	Bone weight	HCW	0.75***	2.124	0.08	0.005	8.99
2		CONF	0.84***	1.727	-0.42	0.110	
3		BCAR	0.85**	1.643	-0.07	0.025	

***$p \leq 0.01$; ***$p \leq 0.001$.*

In the first group of used model (Table 1), which included hot carcass weight and carcass classification as independent variables, explained variation range from 0.93 to 0.97 for muscle, from 0.32 to 0.60 for fat and from 0.76 to 0.85 for bone. When ultrasound measurement were included into the above models (Table 2), it was observed an increase of percentage of variation explained among 1 and 23 percent units for muscle and fat weight, respectively. Including ultrasound measurements, explained variation for bone decreased in a one unit but the RSD reduced by 8.1% as well. Indeed, this decreased in RSD values can be observed for all equations, equally muscle, fat and bone. That is, the accuracy of the predicting equations increased by including ultrasound measurements.

Using HCW and ultrasound measurements, Ripoll *et al.* (2005) found lower percentages of explained variation for muscle fat and bone, when they were converted into carcass percentage. Same authors also increased the explained variations when SEUROP classification was included. HCW was selected in the first place in every model, except for fat weight, being the variable that explains greater percentage of variability, being positively related with the dependent variables. When the dependent variable was fat weight all independents variables, included both ultrasound measured BCAR and CCAR, adjusted better to a logarithmic than to a linear regression.

The independent variables of CONF and BCAR, when are included into the model, were positively related with muscle weight and negatively with fat and bone weight. Conversely, FATNESS and CCAR were positively related with fat and bone weight but negatively with muscle weight.

Conclusions

The residual standard deviation (RSD) values for the regression equations indicate that ultrasonic measurements in addition of hot carcass weight and SEUROP classification system could be acceptable predictors of bull carcass tissue composition.

Acknowledgements

Study funded by the Ministry of Education and Science of Spain and the European Union Regional Development funds (INIA RTA-01-106).

References

Delfa, R., A. Teixeira and C. González, 1992. Composición de la canal. Medida de la composición. En: Calidad de la Canal Ovina (III). Ovis, Volume 23, 9-22.

Delfa, R., 1992. Predicción de la composición corporal y de la canal a partir del animal vivo y de la canal. In: Calidad de la Canal Ovina (III). Ovis, Volume 23, 25-26.

Delfa, R., G. Ripoll, M. Joy, F. Lahoz and P. Albertí, 2005. Predicción del peso de las categorías comerciales del despiece bovino a partir del peso sacrificio, canal caliente, espesores de grasa subcutánea y músculo, tomadas con ultrsonidos en vivo o en la canal y con calibre en la canal. XV Congreso de Zootecnia - I Congreso Iberoamericano de Zootecnia, Vila-Real, España.

Greiner, S.P., G.H. Rouse, D.E. Wilson, L.V. Cundiff and T.L. Wheeler, 2003. The relationship between ultrasound measurements and carcass fat thickness and longissimus muscle area in beef cattle. Journal of Animal Science, Volume 81, 676-682.

Kempster, A.J., A. Cuthertson and G. Harrington, 1982. Carcass evaluation in livestock breeding, production and marketing. Granada Publishing, 306 pp.

Largard, A.G. and K. Kolstad, 2003. Selections of ultrasonic muscle depth; direct and correlated response in a Norwegian experimental sheep flock. Small Ruminant Research 48, 23.29

Ott, R.L., 1993. An introduction to statistical methods and data analysis. 4th ed., Duxbury Press, Belmont, CA, USA, 1051 pp.

Ripoll, G., P. Alberti, F. Lahoz and R. Delfa, 2005. Uso de ultrasonidos en vivo y en la canal, y de la clasificación SEUROP como estimadores de la composición tisular de la canal de bovino. ITEA, Volume 26 (Extra), 729-731.

Texeira, A., S. Matos, S. Rodríguez, R. Delfa and V. Cadavez, 2006. In vivo estimation of lamb carcass composition by real-time ultrasonography. Meat Science, Volume 74, 289-295.

Wallace, M.A, J.R. Stouffer and R.G. Westervelt, 1977. Relationships of ultrasonic and carcass measurements with retail yield in beef cattle. Livestock Production Science, Volume 4, 153-164.

Wilson, D.E., 1992. Application of ultrasound for genetic improvement. Journal of Animal Science, Volume 70, 973-983.

Comparison of two methods for longissimus muscle area measurements

Danijel Karolyi, A. Džidić, K. Salajpal, M. Đikić and I. Jurić
University of Zagreb, Faculty of Agriculture, Animal Science Department, Svetošimunska cesta 25, 10000 Zagreb, Croatia

Abstract

The aim of the study was to determine the variance of *longissimus* muscle area (LMA) measurements associated with the standard use of polar planimeter (REISS Precision 3005, Germany) and Image tool® software program. Measurements were conducted on six dorsal samples taken between 7th and 9th rib of the right half of beef carcasses. Firstly, LMA was traced on acetate paper and measured using a planimeter. Secondly, the images of *longissimus* muscles for computer analysis were obtained by scanning (UMAX PowerLook II) dorsal samples in ratio 1:1. LMA size was determined with Image tool® software. Triplicate measurements of LMA were performed by six measurers for each sample and method. Variance components for LMA measurement method associated with measurer, sample and measurement replicates were estimated by REML. The proportion of the total variance in LMA measurement between samples was greater with Image tool® (99.79%) than with planimeter (96.49%). The proportion of variance between measurers was 0.12% with Image tool® and 0.34% with planimeter. The proportion of error variance due to measurement replicates associated with the use of Image tool® program was smaller (0.09%) than with the planimeter (3.17%). The application of Image tool® showed greater precision in measuring LMA than standard planimetry method.

Keywords: beef cattle, M. longissimus dorsi, planimeter

Introduction

In beef cattle a positive genetic correlation exists between *longissimus* muscle area (LMA) and carcass weight, dressing out percentage and shear of muscle tissue in the carcass (Marshall, 1999). On contrary, there is a negative genetic correlation between LMA and thickness of the fat tissue (Marshall, 1999). Moreover, between LMA and carcass weight or muscle tissue in the carcass a positive phenotyphic correlation exists (Gregory *et al.*, 1995). Measurement of the LMA is used to estimate carcass traits during progeny test in breeding and selection program and to evaluate feeding trials. Standard method used to measure *longissimus* muscle area is to trace by hand LMA on acetate paper. Traced area from the paper is measured using planimetry method and transferred into squared centimetres. The aim of this study was to determine variability of the LMA measurements with a standard planimetry method and by using the Image tool® program for the analysis of the digital picture.

Materials and methods

Longissimus muscle area (LMA) measurements were conducted on six dorsal samples taken between 7th and 9th rib of beef carcasses. LMA was traced on acetate paper and measured using polar planimeter (REISS Precision 3005, Germany). Triplicate measurements of LMA were done by six persons for each sample. Images of muscles for computer analysis were obtained by dorsal samples scanning in ratio 1:1 (UMAX PowerLook II, with resolution 300 dpi, RGB mode). Scanned pictures were processed with the Image tool® program (UTHSCSA Image Tool for Windows version 3.0, The University of Texas Health Science Center, San Antonio, TX, USA), where LMA was measured in resolution 1:2 with calibration line of 588 pixels as 5 cm distance. Triplicate measurements of LMA were performed by six measurers for each sample. Scanned dorsal samples pictures were printed out and traced with planimeter. Triplicate measurements per sample picture were conducted from six measurers. There were total of 54 measures of the *longissimus* muscle area on each dorsal sample. Variance components were estimated by REML method from SAS statistical program (SAS, 1999). The fixed effects in the model were measuring method, while measurer, LMA sample and repeated measurements within interaction (sample x measuring method) were defined as the random effects. Dependent variable was LMA in squared centimetres.

Results and discussion

Descriptive statistics of the *longissimus* muscle area measurements with different measuring methods was shown in Table 1.

LMA in planimetry method measured from the acetate paper was smaller than the LMA measured from the planimetry method measured from the scanned pictures, while the LMA measured with the Imagetool® program was in between. Variance components of *longissimus* muscle area measurements for planimetry and ¨Image tool¨ software were presented in Table 2.

The proportion of error variance due to measurement replicates associated with the use of Image tool program was much smaller (0.09%) than the proportion of error variance due to measurement replicates associated with standard use of planimeter (3.17%). The application of Image tool has 35 time greater precision in *longissimus* muscle area measurement than standard planimetry method. The results of this research show that Image tool® application could be used in LMA measurements.

Table 1. Descriptive statistics of the Longissimus muscle area measurements (cm2) with different measuring methods.

Measuring method	n	Mean	SD	Minimum	Maximum
Planimeter p[1]	108	67.94	9.49	48.00	85.00
Planimeter s[2]	108	72.24	10.70	49.20	89.60
Image tool	108	70.54	11.04	51.41	86.95

[1]Acetate paper.
[2]Scanned dorsal sample of the MLD.

Table 2. Variance components of Longissimus muscle area (cm2) measurements for planimetry and ¨Image tool¨ software.

Measuring method	n	Variance due to			Percentage of the total variance due to		
		Measurer	Sample	Error	Measurer	Sample	Error
Planimeter p[1]	108	0.3619	102.72	3.3711	0.34	96.49	3.17
Planimeter s[2]	108	0.8914	131.32	3.3330	0.66	96.88	2.46
Image tool	108	0.1773	144.66	0.1275	0.12	99.79	0.09

[1]Acetate paper.
[2]Scanned dorsal sample of the MLD.

Conclusions

It is possible to use Image tool® software for LMA measurement. Moreover, application of Image tool® showed greater precision in measuring LMA than standard planimetry method. However, for practical application the usage of digital camera is required.

Acknowledgements

Authors thank to Mr. Dragan Tupajic for his excellent technical assistance and cooperation in experimental part of the research.

References

Gregory, K.E., L.V. Cundiff and R.M. Koch, 1995. Genetic and phenotypic (co)variances for growth and carcass traits of purebred and composite populations of beef cattle. Journal of Animal Science, Volume 73, 1920-1926.

Marshall, D.M., 1999. Genetics of meat quality. In: Fries, R. and A. Ruvinski (eds.), The Genetics of Cattle, CAB International, Wallingford, UK, 605-628.

SAS, 1999. OnlineDoc® Release 8. SAS Institute Inc., Cary, NC, USA.

The use of ultrasound measurements in estimation of valuable cut content in fattened bulls

Peter Polák, J. Daňo, M. Oravcová and E. Krupa
Slovak Agricultural Research Centre, Research Institute for Animal Production, Hlohovská 2, 949 92 Nitra, Slovakia

Abstract

The aim of our investigation was to analyse possibilities of using ultrasound as a method for objectification of price of fattened bulls via estimation of valuable cut content. During the years 2000–2005, 483 bulls of seven breeds (Slovak Spotted, Slovak Pinzgauer, Holstein, Brown Swiss, Simmental and Charolaise) were fattened, measured by ultrasound and slaughtered. Results of detail dissection were used to evaluate the dependence of weight of valuable cut, weight of hot carcass and EUROP grading for carcass conformation obtained by dissection and assessment by ultrasound. The average class for carcass composition of all bulls was 2.706 (R^-). Weight of valuable cats was assessed by linear regression model with weight before slaughter and muscle thickness on the last thoracic vertebra as independent variables. No statistical differences were found out between weights of valuable cut obtained using both methods. Relative value difference was 0.088%, i.e. 123 g of meat. For modelling the dependence of response variable weight of valuable cut on carcass composition, coefficients of non-linear function ($y = 19.5963 + 18.961x - 2.38542\ x^2$; $Iyx = 0.497995$) were estimated. Maximum of this function for carcass composition was $x_{max} = 3.974$ and for weight of valuable cut assessed by ultrasound $y_{max} = 95.74$ kg. Similarly, coefficients of quadratic function ($y = 57.1081 + 44.0243\ x - 5.536796\ x^2$; $Iyx = 0.449525$) were estimated for weight of hot carcass dependent on classification of carcass composition. Maximum of second function for carcass composition was almost the same $x_{max} = 3.98$. Maximum for weight of hot carcass was $y_{max} = 95.74$ kg. Calculated values for extremes of functions for classification of carcass composition (3.974; 3.98) reflect the present structure of the bull's population in Slovakia.

Keywords: ultrasound, beef, carcass composition, valuable cuts

Introduction

The aim of our investigation was to analyse possibilities of using ultrasound (ALOKA SSD-500) for objectification of realisation of fattened young bulls. On the base of our previous results, which were done on 180 carcasses, there existed close dependence among proportion of valuable cut obtained by ultrasound, weight of meat obtained by dissection and weight of hot carcass in absolute and relative numbers as well.

Transfer of grading system of beef carcass to EUROP system in Slovakia and Czech Republic and validation of new norms was signed by subjectivity of classifier. The subjectivity was registered during research and verification of system in practise too. This was the reason to find out new approaches (ultrasound using) in objectification of beef carcass grading.

The hypothesis of utilisation of ultrasound (*in vivo*) as a perspective method of price differentiation and realisation appeared for the first time in the study of Polák and Daňo (2002) where 180 carcasses were used.

Material and methods

Ultrasound measurements were done between 1999 and 2005 according method published by Polák *et al.* (2001) on 183 bulls of seven breeds (Slovak Spotted, Slovak Pinzgauer, Holstein, Brown Swiss, Polled Simmental and Charolaise). Bulls were fattened to the age of 450 to 500 days in three farms.

Weight of valuable cut was assessed by linear regression model where weight before slaughter and muscle thickness on the last thoracic vertebra were used as independent variables. Weight of half carcass, proportion of meat, weight and proportion of valuable cut obtained by ultrasound and by dissection were analysed by non-linear regression analyse. For transcription of alphabetic marks in EUROP grading system numbers were used (5 = E ... 1 = P). Bach and Dunkel (1993) used similar transcription in their work. In the case of inverse transcript (1 = E ... 5 = P) graphic expression of polynomial was inappropriate (Polák and Daňo, 2002).

Results and discussion

Basic statistics of carcass quality variables are in Table 1. Overall average class of carcasses conformation had mean 2.706 (R^-), but proportion of meat was 1% lower in comparison to R class. We think that this fact is validation of existing classifier's subjectivity in carcass classification.

Non-linear regression analysis was used for variables of carcass quality (weight of half carcass, weight of meat obtained by ultrasound and by dissection, proportion of meat obtained by ultrasound and dissection, weight of valuable cut obtained by ultrasound and dissection, proportion of valuable cut obtained by ultrasound and dissection).

Minimal difference (t-test) was found between weight of valuable cut obtained by ultrasound or by dissection. Difference between proportions of valuable cut obtained by dissection or by ultrasound was 0.088%, i.e. 123 g.

After verification of this initial hypothesis the non-linear regression analysis was used to find out dependence between weight of meat obtained by dissection and weight of valuable cut obtained by dissection. Index of correlation was high (I_{xy} = 0.968281) and statistically significant. Similar by

Table 1. Basic statistics of carcass quality variables.

Variables	Conformation of beef carcass (EUROP)					
	E	U	R	O	P	Overall
N	3	75	208	171	26	483
Half carcass (kg)	144.0	143.5	140.1	122.3	97.1	132.0
Meat obtained by dissection (kg)	106.8	106.5	102.5	87.0	68.0	95.8
Meat obtained by ultrasound (kg)	102.1	104.4	103.2	88.4	68.5	96.3
Valuable cut obtained by dissection (kg)	58.3	57.9	55.5	46.6	36.9	51.7
Valuable cut obtained by ultrasound (kg)	56.6	56.7	55.5	47.4	37.3	51.9
Meat obtained by dissection (%)	76.0	75.8	74.9	72.4	71.6	74.0
Meat obtained by ultrasound (%)	77.1	76.4	75.7	71.1	70.3	74.3
Valuable cut obtained by dissection (%)	54.7	50.3	50.3	52.5	54.4	51.3
Valuable cut obtained by ultrasound (%)	55.4	50.2	49.9	52.6	54.4	51.2

high and significant index of correlation was found between weight of half carcass and weight of valuable cut obtained by ultrasound (I_{yx} = 0.972116).

On the base of results of equations mentioned above the classical polynomial of second level was derived and expressed graphically. Dependent variable in the graph was carcass classification (5–1). Independent variables were weight of half carcass (x_1) weight of valuable cut obtained by ultrasound (x_2). Graphical expression of polynomial was very similar to our previous work (Polák and Daňo, 2002).

Because of significant close dependence between dependent variables (weight of meat obtained by dissection – y_1, and weight of half carcass – y_2) on independent one (weight of valuable cut obtained by ultrasound) the dependence of weight of meat obtained by dissection (y_1) and weight of half carcass (y_2) on carcass classification was tested.

Dependence between weight of valuable cut and carcass classification was modelled by quadratic function $y = 19.5963 + 18.961\ x - 2.38542\ x^2$. Maximum of this function was achieved in the point x_{max} = 3.974 (class according to EUROP) with the corresponding weight of valuable cut y_{max} =95.74kg.

Similarly, dependence between weight of half carcass and carcass classification was modelled by quadratic function $y = 57.1081 + 44.0243\ x - 5.536796\ x^2$. Maximum of this function was achieved in the point x_{max} = 3.98 (class according to EUROP) with the corresponding weight of the meat in the carcass side y_{max} = 144.8 kg.

Calculated maximum of the both functions (class U in the EUROP system) support our hypothesis. The results are in agreement with the structure of population of carcass bulls in Slovakia. Estimated functions and observed values of dependence weight of valuable cut obtained by ultrasound and weights of half carcass on classification are reported in Figure 1. Graphical results can be used in the solving of problems in classification of carcasses.

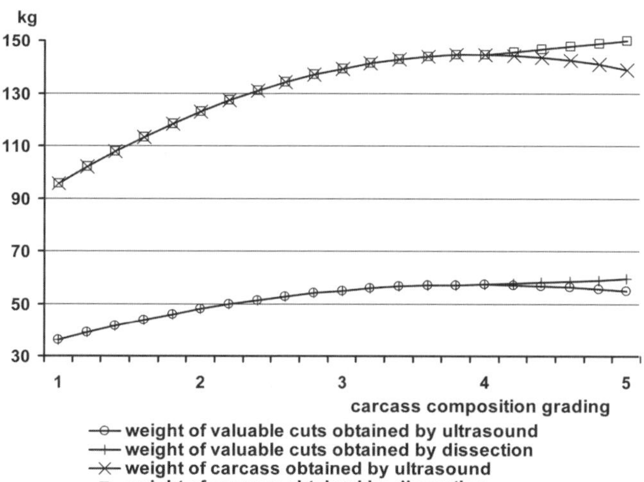

Figure 1. Relationships between of bull's carcass conformation classifications, weight of half carcass and weight of valuable cut obtained by ultrasound.

Conclusion

Majority of young bulls are sold as live animals presently in Slovakia and carcass classification by EUROP grading system does not influence price for bulls but we think, that knowledge obtained in this work will be used in practice in the future. Close dependence between weight of half carcass and weight of valuable cut obtained by ultrasound, and other results too, lead to the conclusion that ultrasound method can be used for classification of carcass conformation and price differential of young bulls in next 10-15 years. More research is needed in this field, however.

Wide use of ultrasound in cattle classification requires cheap ultrasound equipment to by available. Ultrasound could be a cheaper alternative to VIA methods.

References

Bach, H. and R. Dunkel, 1993. Handelklassen für Rindfleisch. Bundesanstalt für Fleischforschun Kulmbach, 24 p.

Polák, P., K. Sloniewski, T. Sakowski, E.N. Blanco Roa, J. Huba and E. Krupa, 2001. In vivo estimates of slauhgter value of bulls using ultrasound and body dimensions. Czech Journal of Animal Science, Volume 46, 159-164.

Polák, P. and J. Daňo, 2002. Prínos sonografie k objektivizácii speňažovania jatočných býkov. Zborník z medzinárodnej vedeckej konferencie. Hovädzí dobytok v novom tisícročí. SPU Nitra, AF, KŠZ, Nitra, 116-123.

The effect of ultrasound probe on accuracy of intramuscular fat content and marbling prediction in beef *longissimus dorsi* muscle

Ján Tomka, P. Polák, E.N. Blanco Roa, E. Krupa, J. Huba, M. Oravcová and D. Peškovičová
Slovak Agricultural Research Centre, Department of Animal Breeding, Hlohovska 2, 949 92 Nitra, Slovak Republic

Abstract

The accuracy of prediction of intramuscular fat content was evaluated using two ultrasound probes (5 MHz/64 mm and 3.5 MHz/172 mm). The study was carried out on 142 bulls (74 Slovak Spotted, 6 Pinzgauer, 22 Holstein, 20 Beef Simmental and 20 Charolais bulls). Ultrasound images of rib eye area were taken between 12th and 13th rib one week before slaughter. After slaughter, photographs of rib eye area were taken in order to determine level of marbling. Computer image analyses using a special software package and statistical analyses using SAS/STAT package were done. No statistically significant differences in level of marbling and intramuscular fat content were found between 5 MHz and 3.5 MHz probes. Correlation coefficients between level of marbling and ultrasound image values were 0.26 (5 MHz) and 0.25 (3.5 MHz). With Slovak Spotted bulls, correlations between SONO and MARB values ranged from 0.52 to 0.69 (3.5 MHz) and from 0.19 to 0.28 (5 MHz probe). The effect of ultrasound probe on intramuscular fat content and marbling prediction accuracy was not shown with whole data set considered. The possible influence of ultrasound probe used is shown with findings for Slovak Spotted bulls considered. The optimal performance for ultrasound machine ALOKA SSD-500 was proposed: 65% (5 MHz probe) and 80 - 90% (3.5 MHz probe).

Keywords: carcass quality, beef cattle, intramuscular fat, ultrasound

Introduction

The frequency of ultrasound probe is one of the main factors influencing ultrasound images. The lower frequency probes are used, the deeper ultrasound penetrations are produced. On the other hand, the images obtained with probes of higher frequencies are more detailed. The aim of this paper was to evaluate the effect of probe frequency on intramuscular fat content and marbling prediction.

Material and methods

The study was carried out on 142 bulls (74 Slovak Spotted, 6 Pinzgauer, 22 Holstein, 20 Beef Simmental and 20 Charolais bulls). ALOKA SSD-500 ultrasound machine with two different ultrasound probes (5 MHz/64 mm and 3.5 MHz/172 mm) was used. Ultrasound images were taken between 12th and 13th rib one week before slaughter. Taking into account previous results, the ultrasound machine performances of 60, 65 and 70% were applied when 5 MHz probe was used (values were entitled as SONO_60, SONO_65 and SONO_70) and performances of 75, 80, 85 and 90% were applied when 3.5 MHz probe was used (values were entitled as SONO_75, SONO_80, SONO_85 and SONO_90). A dissection was performed 24 hours after slaughter. Cross-sectional cut of *musculus longissimus thoracis et lumborum* was performed between 12th and 13th rib. The photographs of cross section area were taken. Values of photograph images were entitled as MARB. Subsequently, the meat samples were taken for laboratory determination of intramuscular fat content using Infratec 1265 Meat Analyzer. Values were entitled as intramuscular fat content (IMF). LUCIA software was used for computer image analyses. SAS 8.2 software was used for statistical analyses. STAT (Basic Statistics) and CORR (Correlation Analysis) procedures were applied.

Results

Basic statistics – whole data set

When 5 MHz probe was used mean SONO values increased along with increasing ultrasound machine performances. The values ranged from 1.60 (SONO_60) to 4.74% (SONO_70) as shown in Table 1. When 3.5 MHz probe was used mean SONO values also increased. The values ranged from 0.58 to 3.98% as shown in Table 2. Mean value of marbling determined from photographic images (MARB) was 2.19%. Mean value of laboratory determined intramuscular fat content (IMF) was 2.39%.

Correlation coefficients – whole data set

The statistically highly significant ($P<0.001$) correlations between single ultrasound measurements were found when 5 MHz or 3.5 MHz probes were used (Tables 3 and 4). When 5 MHz frequency probe was used, the highest correlation between ultrasound measurements and marbling ($r = 0.26$, $P = 0.06$) was found at 65% of overall ultrasound machine performance. When performances of 60 and 70% were applied, correlations between respective ultrasound measurements and MARB values were $r = 0.14$ and were statistically non-significant. Also, no statistically significant correlations were found between SONO values and laboratory determined intramuscular fat content. When 3.5 MHz probe was used, non-significant correlations between SONO and MARB values and between SONO and IMF values were found, except for correlation between SONO_80 and marbling; $r = 0.25$ ($P = 0.09$).

Comparisons – whole data set

The results obtained using different probes were almost similar. Only correlations between single ultrasound measurements were statistically significant. With both probes used, no differences were observed when correlations between SONO values and marbling, and SONO values and laboratory determined intramuscular fat content were compared.

Table 1. Basic statistics (5 MHz probe).

Characteristic	Mean (%)	Standard Deviation
SONO_60	1.60	1.579
SONO_65	2.76	2.583
SONO_70	4.74	3.879
MARB	2.19	0.835
IMF	2.39	1.144

Table 2. Basic statistics (3.5 MHz probe).

Characteristic	Mean (%)	Standard Deviation
SONO_75	0.58	0.64
SONO_80	1.36	1.249
SONO_85	2.24	1.637
SONO_90	3.98	2.284
MARB	2.19	0.835
IMF	2.39	1.144

Table 3. Pearson's correlation coefficients (5 MHz probe).

	SONO_65	SONO_70	MARB	IMF
SONO_60	0.65***	0.72***	0.14	-0.14
SONO_65		0.88***	0.26	-0.07
SONO_70			0.14	-0.18

*** $P < 0.001$ (Fisher's Z-test)

Table 4. Pearson's correlation coefficients (3.5 MHz probe).

	SONO_80	SONO_85	SONO_90	MARB	IMF
SONO_75	0.65***	0.65***	0.55***	0.10	0.04
SONO_80		0.79***	0.64***	0.25	-0.05
SONO_85			0.70***	0.09	-0.08
SONO_90				0.07	-0.19

*** $P < 0.001$ (Fisher's Z-test)

Basic statistics – Slovak Spotted bulls

When data on Slovak Spotted bulls (74 animals) were considered, different results were obtained (Tables 5 and 6). SONO values increased along with increasing ultrasound intensities. Mean value of marbling was 2.46%. Mean value of intramuscular fat content was 2.69%.

Table 5. Basic statistics (5 MHz probe).

Characteristic	Mean (%)	Standard Deviation
SONO_60	1.52	1.455
SONO_65	2.55	2.402
SONO_70	4.27	3.355
MARB	2.46	1.325
IMF	2.69	1.242

Table 6. Basic statistics (3.5 MHz probe).

Characteristic	Mean(%)	Standard Deviation
SONO_75	0.50	0.587
SONO_80	1.09	1.157
SONO_85	1.88	1.417
SONO_90	3.05	2.002
MARB	2.46	1.325
IMF	2.69	1.242

Correlation coefficients – Slovak Spotted bulls

The statistically highly significant correlations ($P<0.001$) between single ultrasound measurements were found when 5 and 3.5 MHz probes were used (Tables 7 and 8). When 5 MHz frequency probe was used, statistically significant correlation between SONO values and marbling ($r = 0.28$; $P<0.05$) was found only at 65% intensity of overall ultrasound machine performance. When 3.5 MHz probe was used, statistically significant correlation between SONO values and marbling ($r = 0.69$; $P<0.05$) was found only at 90% intensity of overall ultrasound machine performance.

Comparisons – Slovak Spotted bulls

With Slovak Spotted bulls, correlations differed when 5 MHz and 3.5 MHz probes were used. Correlations between SONO and MARB values ranged from 0.19 to 0.28 (5 MHz probe) and from 0.52 to 0.69 (3.5 MHz).

Table 7. Pearson's correlation coefficients (5 MHz probe).

	SONO_65	SONO_70	MARB	IMF
SONO_60	0.64^{***}	0.68^{***}	0.19	-0.08
SONO_65		0.87^{***}	0.28^{*}	-0.01
SONO_70			0.19	-0.13

*** $P < 0.001$, * $P < 0.05$ (Fisher's Z-test)

Table 8. Pearson's correlation coefficients (3.5 MHz probe).

	SONO_80	SONO_85	SONO_90	MARB	IMF
SONO_75	0.80^{***}	0.67^{***}	0.66^{***}	0.52	0.17
SONO_80		0.84^{***}	0.65^{***}	0.54	0.11
SONO_85			0.82^{***}	0.60	0.05
SONO_90				0.69^{*}	0.12

Discussion

The effect of ultrasound probe frequency was not shown with whole data set considered. This can be due to the animals (probably of low level of intramuscular fat) used in the experiment. With Slovak Spotted bulls, correlations between ultrasound measurements and marbling, and also, correlations between ultrasound measurements and intramuscular fat content were higher than correlations found with whole data set. In addition, correlations differed with various probes used. This is in contrast with findings based on the whole data set. It is suggested to evaluate the effect of breed on prediction accuracy. Also important is a decision whether direct prediction of intramuscular fat content is efficient, or only prediction of marbling is sufficient. It can be assumed that prediction of marbling is sufficient from the consumer's point of view. For meat quality and palatability, the uniform distribution of intramuscular fat along the whole muscle is needed. Therefore prediction of intramuscular fat content can be considered insufficient.

Conclusions

The hypothesis of influence of ultrasound probe on accuracy of prediction of marbling and intramuscular fat content was not proved. Particular findings for Slovak Spotted bulls, however, indicate the possible influence of ultrasound probe used.

Taking into account the results obtained, it could be said that optimal ultrasound intensity should be applied when particular probe is used in combination with particular ultrasound machine. The optimal performance for ultrasound machine ALOKA SSD–500 was proposed: 65% (5 MHz probe) and 80 - 90% (3.5 MHz probe).

In vivo ultrasonic measurements and live weight for predicting carcass quality in Churra Tensina mountain breed lambs

Rafael Delfa[1], M. Joy[1], A. Sanz[1], B. Panea[1], J. Alvarez-Rodriguez[1], P. Albertí[1] and A. Teixeira[2]
[1]Unidad de Tecnología en Producción Animal, Centro de Investigación y Tecnología Agroalimentaria de Aragón, Avda. Montañana 930, 50016 Zaragoza, España
[2]Departamento de Zootecnia, Escuela Superior Agraria, Apto. 727, 5301-855 Bragança, Portugal

Abstract

Forty-six single male lambs of Churra Tensina mountain breed with a live-weight ranging from 19.9 to 24.4 kg ('Ternasco' carcass commercial type) were scanned by ultrasound (ALOKA model SSD-900, with a 7.5 MHz probe) to determine M. *longissimus dorsi* depth and subcutaneous fat thickness between the 10th-11th, 12th-13th dorsal vertebrae and 1st-2nd, 3rd-4th lumbar vertebrae. Lambs were slaughtered and carcasses were cooled at 4 °C for 24 h and halved. The left side was divided according to a standardised jointing procedure, based on six anatomical regions: shoulder, long leg, anterior ribs, ribs, flank and neck. Each joint was then dissected into muscle, bone plus remainder and subcutaneous, inter-muscular, kidney and pelvic fat. The *in vivo* ultrasound measurements plus slaughter weight were fitted to predict carcass tissue composition by Stepwise regression analysis. All the developed models were highly significant ($P<0.001$) and explained 70, 51, 82, 56, 59 and 41% of the muscle, bone plus remainder, subcutaneous, inter-muscular, kidney and pelvic fat variation respectively. The model residual standard deviations were lower than 124.3 g.

Keywords: lamb, fat thickness, M. longissimus dorsi, ultrasound

Introduction

The EU established a model for classification of lamb carcasses highly subjective, using photographic standards (European Union, 1994). That model has not been useful to classify carcasses according to the most important consumer criteria and it becomes important to provide information about carcass composition as saleable meat, or fat, muscle and bone contents. Therefore, it is necessary to develop an accurate method of carcass classification that allows assessing objectively carcass composition.

Nowadays, one of the most promising techniques to assess carcass composition is Real Time Ultrasound Scanner because it presents a high potential for the immediate use (Delfa *et al.*, 2005; Teixeira *et al.*, 2006). This tool can predict with high precision the tissue and anatomical composition of carcasses in measures, which are carried out *in vivo*. It is evident the interest to obtain equations of prediction for the regional lamb genotypes in order to avoid the errors linked to others genotypes and others weight ranges (Delfa, 1992). Churra Tensina sheep is an endangered local course-wooled hardy breed belonging to Churra group, which is raised for lamb production, although it has been milked for years. The main objective of the present study was to determine the precision of slaughter weight (SW) and ultrasound measures as predictors of carcass composition.

Materials and methods

Forty-six male lambs of Churra Tensina breed, from single birth and unshaven, with live-weight of 22.4±0.96 were used. Twenty four hours before slaughter an ALOKA SSD-900 real time Ultrasound

Scanner was used to take16 measures of subcutaneous fat thickness (C*), 16 of subcutaneous fat thickness plus interface (C**), 16 of subcutaneous fat thickness plus interface plus skin (C), 16 measures of *M. longissimus dorsi* depth (B) and 4 of *M. longissimus dorsi* width (A). These measurements were taken with a 7.5 MHz probe at different anatomical positions (10-11 and 12-13 dorsal vertebrae, Dv; and 1-2, 3-4, lumbar vertebrae, Lv). Probe was placed perpendicularly to the backbone at 2 cm, 4 cm and at 1/3 distance and parallel to the backbone at 1/3 of distance (Figure 1).

Lambs were slaughtered and carcasses were cooled at 4 ºC for 24 h and halved carefully. On the left side, fat thickness was measured with a 'calliper' at the same anatomical point where it was assessed in the live animal using ultrasounds. The left side was divided into six standardized commercial joints, shoulder, long leg, anterior ribs, ribs, flank and neck, according to Colomer-Rocher *et al.* (1988). Depth and width of *M. longissimus dorsi* were taken in the same anatomical localizations as ultrasounds. Each joint was then dissected into muscle, bone plus remainder and subcutaneous, inter-muscular, kidney and pelvic fat, according to Colomer-Rocher *et al.* (1988).

Data were analysed by stepwise regression (Ott, 1993) using the ultrasonic measurements and the SW as independent variables. The statistics R^2 and residual standard deviation (RSD) were used to assess the accuracy of prediction of ultrasonic measurement and SW (Kempster *et al.*, 1982).

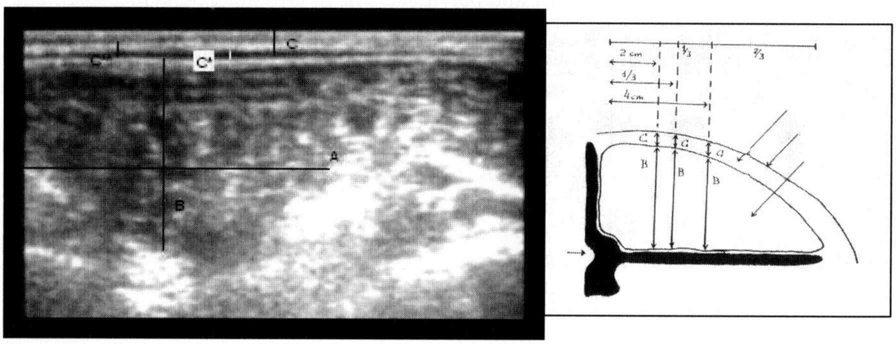

Figure 1. Ultrasounds measurements carried out "in vivo" perpendicularly to the backbone. C= skin+interface+fat; C = fat; C** = interface+fat; B = M. Longissimus dorsi depth; A = M. Longissimus dorsi width. B and C measures were taken at 2cm, 4cm and 1/3 from the backbone.*

Results and discussion

The prediction equations, determination coefficients (R^2) and residual standard deviation (RSD) of the weight of the different carcass tissues are shown in Table 1.

The SW only was admitted as independent variable in the prediction models of muscle, subcutaneous and kidney fat weights. These results agree with those found by Cadavez *et al.* (2000b) and Joy *et al.* (2005), but are not supported by those found by Delfa *et al.* (1995), Teixeira and Delfa (1997), Cadavez *et al.* (2000a) and Teixeira *et al.* (2006) where live-weight was the first variable admitted in the models. These differences can be due to the different type of animal and the range of live-weight and the anatomical points used for ultrasonic measurements.

All prediction equations included different muscle depth (B) measurements and subcutaneous fat thickness, mainly the C measure. The prediction equation of bone weight was the only that include

Table 1. Multiple regression equations live-weight and ultrasound measurements to predict carcass composition. (Linear equation: y=a+bx; Dep. Var. (y) = Dependent variable, Independent Var. (x) = Independent variable a= Intercept, b=slope).

Step	Dep. Var. (y)	Independent Var. (x)	R^2	RSD	b	a
1	Muscle weight ½	B parallel 1/3 1st- 2nd Lv	0.61***	137.48	52.42	247.28
2	carcass	SW	0.66***	130.46	62.18	
3		B 1/3 1st- 2nd Lv	0.70***	124.31	26.71	
1	Bone plus	B parallel 1/3 1st- 2nd Lv	0.20*	72.70	4.33	344.1
2	remainder weight	A 12th- 13th Dv	0.39***	64.28	6.34	
3	½ carcass	B 2 cm 12th- 13th Dv	0.46***	61.40	15.52	
4		C* 4 cm 3rd- 4th Lv	0.51***	59.14	-47.65	
1	Subcutaneous fat	C 1/3 1st- 2nd Lv	0.57***	59.14	64.28	-608.12
2	weight ½ carcass	C 4 cm 10th- 11th Dv	0.69***	51.12	27.97	
3		SW	0.76***	44.98	23.51	
4		B 4 cm 3rd- 4th Lv	0.79***	42.32	9.74	
5		C parallel 1/3 12th- 13th Dv	0.82***	40.57	-20.70	
1	Intermuscular fat	B 4 cm 3rd- 4th Lv	0.28**	99.25	29.85	-47.21
2	weight ½ carcass	C 2 cm 1st- 2nd Lv	0.43***	89.65	99.24	
3		B parallel 1/3 1st- 2nd Lv	0.51***	84.07	-19.08	
4		B parallel 1/3 3rd- 4th Lv	0.56***	80.44	12.06	
1	Kidney fat weight	SW	0.30**	27.01	13.04	-216.69
2	½ carcass	B 2 cm 3rd- 4th LV	0.42***	24.85	7.66	
3		B 2 cm 1st- 2nd Lv	0.49***	23.50	-6.61	
4		C* parallel 1/3 10th- 11th Dv	0.55***	22.27	8.54	
5		B 4 cm 1st- 2nd Lv	0.59***	21.58	2.85	
1	Pelvic fat weight ½	B 4 cm 3rd- 4th Lv	0.12*	10.27	3.89	35.72
2	carcass	B 2 cm 12th- 13th Dv	0.25**	9.56	-3.47	
3		C 4 cm 3rd- 4th Lv	0.36**	8.96	8.67	
4		C parallel 1/3 1st- 2nd Lv	0.41***	8.72	4.69	

A measurement equation with A measurement as independent variable and when associated with a B measurement increase the accuracy of the model in 19%. When the SW was included in the equations of ULTRASOUNDS measures the accuracy of the predicted weight of the different carcass tissues (R^2, RSD) was enhanced between 9 and 31%. All the developed prediction equations were highly significant ($P<0.001$) and explained 70, 51, 82, 56, 59 and 41 % of the muscle, bone plus remainder, subcutaneous, intermuscular, kidney and pelvic fat variation, respectively. All the RSD of the predicting equations were very low and the highest error was only 124.3 g to predict the muscle carcass weight.

The 70% of variation in muscle weight was accounted for by variation in measurements B parallel 1/3, 1st-2nd Lv, SW and B1/3, 1st-2nd Lv, in agreement with Cadavez *et al.* (2000b) and Joy *et al.* (2005). However, Teixeira and Delfa (1997) and Teixeira *et al.* (2006) found greater R^2 values (0.94 and 0.96, respectively).

The B, A and C* measurements together (Table 1) accounted for the 51% of variation in bone weight. These results were very similar to those reported by Delfa *et al.* (1995). In relation to the subcutaneous fat weight, we observed that 57% of the variations were accounted for by variation in C1/3, 1st-2nd Lv. The inclusion of the measurement C, SW and B (Table 1) increased 25% the determination coefficient and decreased 31.4% RSD, which is in agreement with values observed by Teixeira and Delfa (1997), Cadavez *et al.* (2000b), Joy *et al.* (2005), and Teixeira *et al.* (2006).

Nevertheless, only 56% of variation in intermuscular fat weight was accounted for by variation in B and C measurements (Table 1). Nevertheless Joy *et al.* (2005) working with lambs of the same breed and with similar live-weight range, reported that 75% of variation in the weight of these depots were accounted for by variation in three C measurements.

Finally, for kidney and pelvic fat we observed that the B and C measurements and SW explained 59 and 41% respectively of the variation in the weight of this fat depots (Table 1), according with the results found by Joy *et al.* (2005).

Conclusions

From the present animal type it was no advisable to take subcutaneous fat thickness and subcutaneous fat thickness plus interface measures measurements to predict carcass composition. The low residual standard deviation in association with the good determination coefficients indicate that ultrasound measurements in addition to slaughter-weight could be acceptable predictors of some carcass tissue components as muscle weight and subcutaneous fat weight in Churra-Tensina lambs.

Acknowledgements

Study funded by the Ministry of Education and Science of Spain and the European Union Regional Development funds (INIA RTA-03-031).

References

Cadavez, V., A. Teixeira, R. Delfa and S. Matos, 2000a. Prediction of carcass composition in vivo by slaughter weight and ultrasound measurements in Churro Galego Bragançano local breed lambs. Abstracts 51st Annual Meeting EAAP, 296.

Cadavez, V., A. Teixeira, R. Delfa and S. Rodríguez, 2000b. Utilización de ultrasonidos y el peso vivo para la predicción in vivo de la composición de la canal de corderos. Producción Ovina y Caprina SEOC, Núm. XXV, 165-168.

Colomer-Rocher, F., P. Morand-Fehr, A.H. Kirton, R. Delfa and I. Sierra-Alfranca, 1988. Métodos normalizados para el estudio de los caracteres cuantitativos y cualitativos de las canales caprinas y ovinas. Cuadernos INIA, Volume 17, 41 pp.

Delf, R., 1992. Predicción de la composición corporal y de la canal a partir del animal vivo y de la canal. In: Calidad de la canal ovina. (III). Ovis, 23, 25-56.

Delfa, R., A. Teixeira, C. González and I. Blasco, 1995. Ultrasonic estimates of fat thickness and longissimus dorsi muscle depth for predicting carcass composition of live Aragón lambs. Small Ruminant Research, Volume 16, 159-164.

Delfa, R., A. Teixeira, V. Cadavez and I Sierra-Alfranca, 2005. Predicción in vivo de la composición de la canal: técnica de los ultrasonidos y puntuación de la condición corporal. In: Estandarización de las metodologías para evaluar la calidad del producto (animal vivo, canal, carne y grasa) en los rumiantes. Monografías INIA - Serie Ganadera nº 3, 61-87.

European Union, 1994. Council Regulation Nº 1278/94 of 30 May 1994 modifying council Regulation No, 2137/92, determining the Community scale for the classification of ovine carcasses. D.O.C.E. no L140, p. 5.

Joy, M., M. Oloriz, A. Sanz, D. Villalba, R. Revilla and R. Delfa, 2005. Utilisation de mesures par ultrasons pour la prédiction de la composition tissulaire de la carcasse chez des agneaux vivants de race Churra Tensina. 12ª Recontres Recherches Rumiants, Paris, France, 397.

Kempster, A.J., A. Cuthertson and G. Harrington, 1982. Carcass evaluation in livestock breeding, production and marketing. Granada Publishing, London, UK, 306 pp.

Ott, 1993. An introduction to statistical methods and data analysis. 4th ed., Duxbury Press, Belmont, CA, USA, 1051 pp.
Teixeira, A. and R. Delfa, 1997. The use of ultrasonic measurements assessed with two probes in live lambs for prediction the carcass composition. Abstracts 48th Annual Meeting EAAP, 295.
Texeira, A., S. Matos, S. Rodríguez, R. Delfa and V. Cadavez, 2006. In vivo estimation of lamb carcass composition by real-time ultrasonography. Meat Science, Volume 74, 289-295.

Longissimus thoracis et lumborum muscle volume calculation using *in vivo* real time ultrasonography

Severiano R. Silva, C.M. Guedes, V.A. Santos, A.L. Lourenço, J.M.T. Azevedo and A. Dias-da-Silva
CECAV-Universidade de Trás-os-Montes e Alto Douro, Department of Animal Science, Apartado 1013, 5000-801 Vila Real, Portugal

Abstract

The *Longissimus thoracis et lumborum muscle* (LM) volume was measured *in vivo* by real-time ultrasonography (RTU) in 13 female sheep. Animals were scanned over 6 identified sites (7th, 9th, 11th and 13th thoracic vertebrae and 2nd and 4th lumbar vertebrae). After RTU measurements, the animals were weighed and slaughtered. RTU volume measurements were calculated by multiplying the LM area obtained at each site by the slice lengths. Equivalent measurements to those taken *in vivo* were obtained on the carcass using a digital camera and image analysis software. Correlation analysis was carried out to evaluate relationships between LM volume measured *in vivo* by RTU and in carcass. The LM volume measured in carcass and *in vivo* by RTU was highly correlated (r= 0.96, 0.96 and 0.98 for lumbar, thoracic and all vertebrae, respectively). These results strongly support that LM volume can be predicted *in vivo* by RTU.

Keywords: longissimus muscle, sheep, volume, ultrasound

Introduction

The determination of three-dimensional structure from serial sections is a common problem in animal corporal composition studies. Computer tomography (CT) and magnetic resonance image (MRI) analysis are frequently used for *in vivo* three-dimensional structure determination by using consecutive images that can be reconstructed to render a volume measurement of a region of interest (Szabo *et al.*, 1999; Mitchell *et al.*, 2001). Studies conducted with pigs (Baulain, 1997; Mitchell *et al.*, 2001) and with lambs (Jopson *et al.*, 1995; Kvame and Vangen, 2006) showed that volume measured *in vivo* by CT and MRI was highly correlated (r > 0.90) with the weights of the dissected tissues. Thus, these techniques have been pointed out as very accurate but its high cost limits the routinely application in animal science (Fuller *et al.*, 1994). The real time ultrasonography (RTU) can also be used for volume determinations.

Since almost five decades, ultrasounds have been used for evaluating animal composition based on tissue depth and area measurements (Stouffer *et al.*, 1961; McEwan *et al.*, 1989; Hopkins *et al.*, 1993; Silva *et al.*, 2005; Teixeira *et al.*, 2006). Compared to tissue area and depth, much less information is available about volume measurements. Results obtained with lamb (Maghoub, 1998) and with broilers (Silva *et al.*, 2006a) showed that muscle volume measurements obtained *in vivo* by RTU were able to explain carcass composition. Maghoub (1998) found that *longissimus thoracis et lumborum* muscle (LM) volume measured *in vivo* by RTU was correlated (r = 0.59; n = 18) with LM volume determined on carcass.

The study herein reported was undertaken to evaluate the capacity of RTU to measure *in vivo* LM volume in female sheep.

Materials and methods

The experimental group consisted of 13 female sheep from the Île-de-France breed (59.3 ± 8.9 kg) selected from the research herd of the University of Trás-os-Montes and Alto Douro (Vila Real, Portugal). All animals were kept under the same management conditions and were fed according to the AFRC (1993) recommendations. Animal handling followed the EU directive number 86/609/EEC concerning animal care. Prior to ultrasound measurements and subsequent slaughter, animals were shorn and deprived of food for 12 h.

Ultrasound measurements

Just before slaughter, animals were scanned with an Aloka SSD 500V real time scanner using a linear probe of 5.0 MHz (UST-588U-5, 64 mm, Tokyo, Japan). The wool at each measurement point was clipped close to the skin and shaved and a medical ultrasound gel was used as a coupling medium. The probe was placed perpendicular to the backbone over the following sites: 7th, 9th, 11th and 13th thoracic vertebrae and 2nd and 4th lumbar vertebrae, where RTU images were taken (Figure 1a). During ultrasound scanning, sheep were individually restrained. Once a satisfactory image had been obtained at each site, it was captured on a video printer for image analysis.

Slaughter procedure and carcass measurements

After RTU measurements, the live weight of all animals was recorded. All the animals were stunned with a captive bolt gun and slaughtered by exsanguination. The fore and the hind limbs (feet) were then separated at the radio-carpal and tarso-metatarsal articulations, respectively and the pelt, head and all internal organs were removed. After being stored at 4 °C for 24 h, carcasses were weighed and split down by the vertebral column with a band saw. A segment of the thoracic/lumbar region (6th thoracic vertebra to the 5th lumbar vertebra) was removed from the right half to take carcass measurements equivalent to those taken *in vivo* by RTU. This segment was frozen over a horizontal surface in order to minimize muscle shape deformation. Then this segment was split by each vertebra with an electric saw and the planes where RTU measurements were taken were exposed (Figure 1b). After that, it was used a digital camera to capture images of the planes where the carcass measurements were taken and performed image analysis.

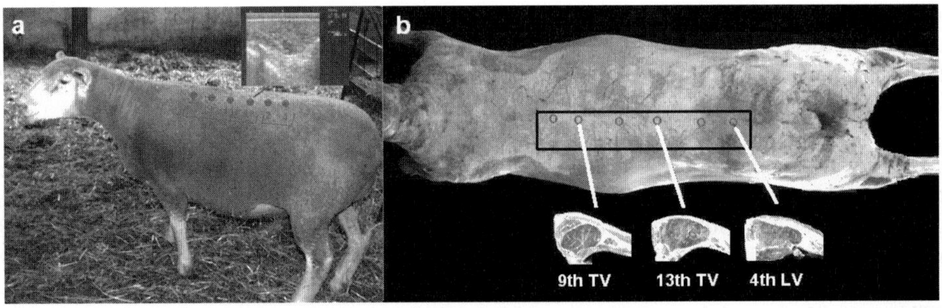

Figure 1. (a) Measurements points for RTU. Shown example for RTU image over the 13th thoracic vertebra. (b) carcass measurements points and joint examples for 9th, 13th thoracic vertebrae (TV) and 4th lumbar vertebra (LV) showing the planes where the LM area was taken.

Image analysis and volume calculation

The printed images were digitized and RTU measurements taken. The area of LM for both RTU and carcass images, at each site, were determined after image analysis using the Image J software (http://rsb.info.nih.gov/ij/). The area was obtained by tracing the contour of LM and by counting the number of pixels on each of the 6 sites. The number of pixels was then converted into area measurements corresponding to the image plane of the 6 mentioned sites. Six areas one for each site was obtained.

The LM volume (LMV) was calculated by multiplying the areas obtained by *in vivo* RTU and carcass measurements by the slice lengths. The slice lengths were obtained after physical measurement *in vivo* or in carcass, using a ruler, of the length between the 6th thoracic vertebra and the 5th lumbar vertebra and divided by 6 to obtain the length of each slice. The following equation was used for LM volume calculation (cm^3):

$$\text{Volume} = \sum_{i=1}^{6} A_i d_i$$

where: d - is the slice length (cm); A - is the site area (cm^2); i - is the number of slices.

The LM total volume (LMVtot) was calculated as the sum of the volumes of 6 slices: LMV7, LMV9, LMV11, LMV13, LMV2 and LMV4. Volumes of the LM from the 7th to the 13th thoracic vertebrae (LMV7–13) and from the 2nd to the 4th lumbar vertebrae (LMV2–4) were also calculated.

Statistical analysis

Data were subjected to correlation analysis to study relationships between carcass and RTU measurements. All analyses were performed with SAS software (v. 8.2; SAS Institute Inc. Cary, NC).

Results and discussion

Live and carcass weight

Mean values, range of the values, standard error (SE) and coefficient of variation (CV) for live weight (LW) and carcass weight (CW) are presented in Table 1. These data show that the sheep presented a large range of variation in LW and CW.

Table 1. Mean values, range of values, standard error (SE) and coefficient of variation (CV, %) for live weight (LW) and carcass weight (CW).

	Mean	Range	SE	CV
Live weight (kg)	59.3	47.1-74.5	2.46	29.3
Carcass weight (kg)	31.2	26.1-40.6	1.31	15.1

Carcass and ultrasound measurements

Mean values, SE and CV for carcass and RTU measurements and the simple correlation coefficients (*r*) between RTU and carcass LM volume are presented in Table 2. The variation observed for all measurements is large. The LM volume increased from the 7th thoracic vertebra to the 4th lumbar vertebra for both carcass and RTU measurements. This finding was expected and it is, obviously, related to the LM shape (Korn *et al.*, 2005).

The LM volume measured *in vivo* by RTU and in carcass were different; the RTU measurements were lower than the carcass measurements. Similar trends were also reported in sheep for LM volume by Mahgoub (1998) and for LM area by Hamby *et al.* (1986), Edwards *et al.* (1989), Ward *et al.* (1992) and Férnandez *et al.* (1998). This is also in agreement with results for LM area found in other animal species (cattle, Greiner *et al.*, 2003; poultry, Silva *et al.*, 2006a). The reasons generally accepted for this underestimation of carcass measurements by RTU are associated with difficulties on image analysis, differences between operators, and differences in muscle shapes due to slaughter procedure. In the present study all the measurements were done by the same operator and all the images were analysed using a computer program. So, it is reasonable to accept that differences in muscle shapes due to slaughter procedures were responsible for the underestimation of carcass measurements.

The ability of RTU for predicting carcass LM volume was high (r varied between 0.703 and 0.981; P<0.01). The lower correlation coefficients were found when muscle measurements were taken over the 7th and the 9th thoracic vertebrae. In these sites the LM is inside a more complex tissue distribution compared with the other sites along the thoracic-lumbar axis (Simm, 1983; Russel, 1995). This anatomical difference and the low absolute values of muscle measurements taken over the 7th and the 9th thoracic vertebrae are the possible reasons that explain these results. The image analysis

Table 2. Mean values, standard error (SE) and coefficient of variation (CV, %) for Longissimus thoracis et lumborum muscle volume (LMV) measured in vivo by real-time ultrasonography (RTU) and in carcass of sheep (n = 13) and the simple correlation coefficients (r) between in vivo real-time ultrasonography (RTU) and carcass measurements.

LMV (cm^3)	Carcass			*In vivo* RTU			*r*
	Mean	SE	CV	Mean	SE	CV	
LMV7[a]	63.7	4.5	25.6	61.1	4.1	24.1	0.813
LMV9[a]	86.7	3.4	14.1	82.6	3.8	16.6	0.703
LMV11[a]	101.9	5.0	17.7	100.9	5.1	18.1	0.942
LMV13[a]	130.5	7.2	19.8	127.9	6.6	18.5	0.935
LMV7-13[b]	382.8	18.3	17.2	372.4	18.6	18.0	0.958
LMV2[a]	138.4	6.1	16.0	127.0	5.6	15.8	0.921
LMV4[a]	119.5	12.5	37.8	112.4	9.0	29.0	0.951
LMV2-4[c]	257.9	17.9	25.1	239.3	13.9	21.1	0.957
LMVtot[d]	640.7	35.6	20.0	611.8	32.0	18.9	0.981

[a] LMV7, LMV9, LMV11, LMV13, LMV2, LMV4: *Longissimus thoracis et lumborum* muscle volume above the 7th, 9th, 11th and 13th thoracic vertebrae and the 2nd and 4th lumbar vertebrae, respectively.
[b] LMV7-13: *Longissimus thoracis et lumborum* muscle volume between the 7th and the 13th thoracic vertebrae.
[c] LMV2-4: *Longissimus thoracis et lumborum* muscle volume between the 2nd and the 4th lumbar vertebrae.
[d] LMVtot: *Longissimus thoracis et lumborum* muscle total volume.
All correlation coefficients (r) are significant *** $P < 0.01$.

using a computer program combined with the ultrasound equipment used in this study, namely the probe length (64 mm), were able to accurately measure the LM volume, allowing the assessment of differences between animals. Improvement in the predicting ability of RTU can be achieved by using large probes and image analysis (McLaren *et al.*, 1991; Young and Deaker, 1994; Williams, 2002; Silva *et al.*, 2006b). The length of the probe may have contributed to the good prediction ability of LM volume in our study because it allows the identification of the lateral boundaries of the LM on RTU images, as observed by Stouffer (2004) in cattle and Silva *et al.* (2006b) in sheep.

Few studies (Mahgoub, 1998; Silva *et al.*, 2006a) have been published on LM volume determinations *in vivo* by RTU. Mahgoub (1998) reported a correlation coefficient of 0.59 ($P<0.01$) between LM volume measured *in vivo* by RTU and in carcass, in sheep ($n = 18$), and Silva *et al.* (2006a) obtained a correlation coefficient of 0.866 ($P<0.01$) between breast volume measured *in vivo* by RTU and in carcass, in broilers ($n = 103$). In the present study, correlation coefficients obtained between LM volume determined *in vivo* by RTU and in carcass were higher ($r = 0.968$; $P<0.01$ for LMVtot) than those obtained by these authors. Several authors determined LM volume using CT (Jopson *et al.*, 1995; Kvame and Vangen, 2006) and found high correlations between *in vivo* and carcass measurements in sheep ($r > 0.90$). Findings of the current study show that LM volume measured *in vivo* by RTU is able to accurately predict carcass LM volume in sheep.

Conclusions

This study showed that LM volume measurements taken *in vivo* by RTU are highly correlated to corresponding carcass measurements. Results issued from this study encourage the use of LM volume measured by *in vivo* RTU. This approach is non-invasive, accurate, reliable, and easy to use and require an inexpensive common RTU machine. Moreover, such measurements in everyday practice are easy to obtain due to the equipment mobility.

References

AFRC Agricultural and Food Research Council, 1993. Energy and protein requirements of ruminants. An advisory manual prepared by the AFRC technical Committee on response to Nutrients, CAB International, Wellingford, UK.

Baulain, U., 1997. Magnetic resonance imaging for the *in vivo* determination of body composition in animal science. Computers and Electronic in Agriculture, Volume 17,189-203.

Edwards, J.W., R.C. Cannell, R.P. Garrett, J.W. Savell, H.R. Cross and M.T. Longnecker, 1989. Using ultrasound, linear measurements and live fat thickness estimates to determine the carcass composition of market lambs. Journal of Animal Science, Volume 67, 3322-3330.

Férnandez, C.A. García, H. Vergara and L. Gallego, 1998. Using ultrasound to determine fat thickness and *longissimus dorsi* area on Manchego lambs of different live weight. Small Ruminant Research, Volume 27, 159-165.

Fuller, M.F., P.A. Fowler, G. McNeill and M.A. Foster, 1994. Imaging techniques for the assessment of body composition. Journal of Nutrition, Volume 124, 1546S-1550S.

Greiner, S.P., G.H. Rouse, D.E. Wilson, L.V. Cundiff and T.L. Wheeler, 2003. The relationship between ultrasound measurements and carcass fat thickness and *longissimus* muscle area in beef cattle. Journal of Animal Science, Volume 81, 676-682.

Hamby, P.L., J.R. Stouffer and S.B. Smith, 1986. Muscle metabolism and real-time ultrasound measurement of muscle and subcutaneous adipose tissue growth in lambs fed diets containing a beta-agonist. Journal of Animal Science, Volume 63, 1410-1417.

Hopkins, D.L., K.L. Pirlot, A.H.K. Roberts and A.S. Beattie, 1993. Changes in fat depths and muscle dimensions in growing lambs as measured by real-time ultrasound. Australian Journal of Experimental Agriculture, Volume 33, 707-712.

Jopson, N.B., K. Kolstad, E. Sehested and O. Vangen, 1995. Computed tomography as an accurate and cost effective alternative to carcass dissection. Proceeding Australian Association Animal Breeding and Genetics, Volume 11, 635-638.

Korn, S. V., U. Baulain, M. Arnold and W. Brade, 2005. Nutzung von magnet-resonanz-tomographie und ultraschall-technik zur bestimmung des schlachkörperwertes beim schaf. Zuchtungskunde, Volume 77, 382-393.

Kvame, T. and O. Vangen, 2006. *In-vivo* composition of carcass regions in lambs of two genetic lines, and selection of CT positions for estimation of each region. Small Ruminant Research, Volume 66, 201-208.

Mahgoub, O., 1998. Ultrasonic scanning measurements of the *Longissimus thoracis et lumborum* muscle to predict carcass muscle content in sheep. Meat Science, Volume 48, 41-48.

McEwan, J.C., J.N. Clarke, M.A., Knowler and M. Wheeler, 1989. Ultrasonic fat depths in Romney lambs and hoggets from lines selected for different production traits. Proceeding New Zealand Society Animal Production, Volume 49, 113-119.

McLaren, D.G., J. Novakofski, D.F. Parrett, L.L. Lo, S.D. Singh, K.R. Neumann and F.K. McKeith, 1991. A study of operator effects on ultrasonic measurements of fat and *longissimus* muscle area in cattle, sheep and pigs. Journal of Animal Science, 69, 54-66.

Mitchell, A.D., A.M. Scholz, P.C. Wange and H. Song, 2001. Body composition analysis of the pig by magnetic resonance imaging. Journal of Animal Science, Volume 79, 1800-1813.

Russel, A.J.F., 1995. Ultrasonography and body composition in sheep. In: Goddard, P.J. (ed.), Veterinary Ultrasonography. CAB International, Wallingford, UK, 315-323.

Silva, S.R., M.J. Gomes, A. Dias-da-Silva, L.F. Gil and J.M.T. Azevedo, 2005. Estimation *in vivo* of the body and carcass chemical composition of growing lambs by real-time ultrasonography. Journal of Animal Science, Volume 83, 350-357.

Silva, S.R., V. Pinheiro, C.M. Guedes and J.L. Mourão, 2006a. Prediction of carcass and breast weights and yields in broiler chickens using breast volume determined *in vivo* by real-time ultrasonic measurement. British Poultry Science, Volume 47, 694-699.

Silva, S.R., J.J. Afonso, V.A. Santos, A. Monteiro, C.M. Guedes, J.M.T. Azevedo and A. Dias-da-Silva, 2006b. *In vivo* estimation of sheep carcass composition using real time ultrasound with two probes of 5 and 7.5 MHz and image analysis. Journal of Animal Science, Volume 84, 3433-3439.

Simm, G., 1983. The use of ultrasound to predict the carcass composition of live cattle – a review. Animal Breeding Abstracts, Volume 51, 853-875.

Stouffer, J.R., 2004. History of ultrasound in animal science. Journal of Ultrasound Medicine, Volume 23, 577-584.

Stouffer, J.R., M.V. Wallentine, G.H. Wellington and A. Diekmann, 1961. Development and application of ultrasonic methods for measuring fat thickness and rib-eye area in cattle and hogs. Journal of Animal Science, Volume 20, 759-767.

Szabo, Cs., L. Babinszky, M.W.A. Verstegen, O. Vangen, A.J.M. Jansman and E. Kanis, 1999. The application of digital imaging techniques in the *in vivo* estimation of body composition of pigs: a review. Livestock Production Science, Volume 60, 1-11.

Teixeira, A., S. Matos, S. Rodrigues, R. Delfa and V. Cadavez, 2006. *In vivo* estimation of lamb carcass composition by real-time ultrasonography. Meat Science, Volume 74, 289-295.

Ward, B.G., R.W. Purchas and A.Y. Abdullah, 1992. The value of ultrasound in assessing the leg muscling of lambs. Proceeding New Zealand Society Animal Production, Volume 52, 33-36.

Williams, A.R., 2002. Ultrasound applications in beef cattle carcass research and management. Journal of Animal Science, Volume 80, Suppl. 2, E183-E188.

Young, M.J. and J.M. Deaker, 1994. Ultrasound measurements predict estimate adipose and muscle weights better than carcass measurements. Proceeding New Zealand Society Animal Production, Volume 54, 215-217.

Slaughter value evaluation of large weight Ile de France and Hungarian Merino lambs by CT and traditional slaughter cutting

Gyula Toldi[1], A. Molnár[2], T. Németh[2] and S. Kukovics[2]
[1]Kaposvár University, Faculty of Animal Science, Guba Sándor u. 1., 7400 Kaposvár, Hungary
[2]Research Institute for Animal Breeding and Nutrition, Gesztenyés u. 1., 2053 Herceghalom, Hungary

Abstract

The slaughter value of two sheep breeds, Hungarian Merino (HM) and Ile de France (IdF) has been compared, by investigation of ten lambs of each genotype and sex on the day before slaughter. CT-images were taken from individuals from occiput to knee. The individuals were slaughtered at ~30 kg live weight and categorized according to the S/EUROP carcass qualification method. After 24h cooling the cut halves were divided according to the slaughterhouse practice applied in Hungary, then boned. The divided parts were classified to 1st class (roast meat) and 2nd class (non-roast meat) quality categories. Close correlations were stated in both genotypes between the tissue (fat, muscle, bone) area results calculated from the CT-images and the weight data measured after slaughter. There was also a favourable correlation between the numbered „body conformation' values of the S/EUROP cut body, and the measured slaughterhouse cutting and boning values. Cut halves of the IdF lambs were superior to the HM lambs mainly in the pieces of roast meat category. The muscle tissue areas of the IdF lambs were always superior as compared to HM lambs of the same live weight. A bit higher fat incorporation was verified in the IdF individuals by both the CT-images and slaughterhouse evaluation.

Keywords: sheep, carcass traditional classification, S/EUROP system

Introduction

The Ile de France breed is a meat – wool type sheep. Its development started in France some 170 years ago (Bajadoux, 1954; Charlet and Bougler, 1981; Bouix *et al.*; 1986; Bouix, 1988), and its pedigree breeding began in 1932. This breed firstly appeared in Hungary in late 1960 and early 1970, and the pedigree breeding launched in early 1980's (Kukovics, 2000). Meat and carcass quality of various genotypes of sheep were tested during the last ten years in Hungary (Mezőszentgyörgyi, 2000; Rózsáné Várszegi, 2003). As the S/EUROP system became official it seemed that the old carcass grading system might be more advantageous (Toldi, 2003) for the Hungarian sheep.

The aim of our present study was to compare the slaughter value of Hungarian Merino and Ile de France fattened lambs using the old carcass grading system and the S/EUROP classification method.

Material and methods

Ile de France and Hungarian Merino lambs were fattened up to 30 kg body weight. Ten lambs from each breed (50-50% male and female) were investigated by CT on the day before slaughter. CT images were taken and recorded from each individual from muscle tissue, fat deposit, bone tissue and tissues with water density. These records were made longitudinally by the use of a whole frame of records of carcass.

All twenty lambs were slaughtered at ~30 kg live weight and categorized according to the S/EUROP sheep carcass qualification system, obligatory in the European Union member states. The S/EUROP carcasses grading were carried out by the same professional person in all cases. Letter signs (S, E, U, R, O, P) of the classification system were changed into numeral codes (Table 1 and 2) to be able for statistical analysis.

After classification the carcass was cut using the butchers' cutting system shown on Figure 1. Thigh, short and long chops were regarded as first class parts (roast meat), while neck, thin flank and shoulder were ranked in the second class meat parts (non-roast meat).

Computed medical imaging (Hounsfield, 1980) and PC based postprocessing software (Berényi-Kövér, 1991), and SPSS for Windows 10.0 software were used in processing the data.

Table 1. Body conformation classes of the S/EUROP system and the applied codes.

S/EUROP	P			O			R			U			E			S
	-	0	+	-	0	+	-	0	+	-	0	+	-	0	+	0
value	1	2	3	4	5	6	7	8	9	10	11	12	13	14	15	16

Table 2. Fat deposit classes of the S/EUROP system and the applied codes.

S/EUROP	1			2			3			4			5		
	-	0	+	-	0	+	-	0	+	-	0	+	-	0	+
value	1	2	3	4	5	6	7	8	9	10	11	12	13	14	15

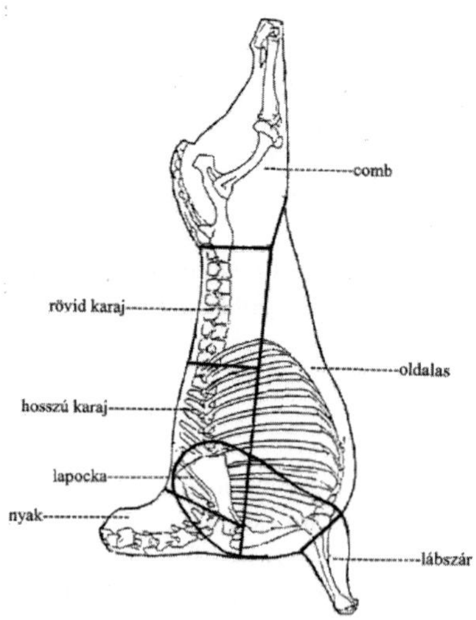

Figure 1. Draft of official butchers' cuts (neck, shoulder, long chop, short chop, leg, ribs, shank).

Results

CT analysis

The area of muscles and fat deposits were bigger in Ile de France than in Merino lambs, however, this difference was not significant. Ile de France had significantly ($P \leq 0.01$) better results in the muscle and fat deposit area of the carcass comparing to the trunk length (Table 3).

In order to evaluate the degree of fat deposition a calculated value was created as a model, like: fat deposit area of the carcass divided by the carcass weight and the fat deposit area as the ratio of the whole areas of carcass. In both cases Ile de France lambs produced higher results in fatness, however, these difference were not proved to be significant.

S/EUROP classification

Body conformation of animals belonging to the two studied breeds differed significantly ($P \leq 0.001$) with the benefit of Ile de France breed. Abdominal and kidney fat deposition had lower degree in Ile de France, but this difference was not significant (Table 3).

Table 3. Statistical comparison of the parameters investigated.

Parameter	Breed	n	$\bar{x}$	±s.d.	P-value[1]
Fat deposit area of carcass by CT (cm^2)	Ile de France	10	1648.40	377.20	0.0532
	Hung. Merino	10	1275.90	426.31	
Muscle tissue area on CT-segments (cm^2)	Ile de France	10	4275.40	257.95	0.2840
	Hung. Merino	10	4076.60	494.28	
Muscle + fat area on CT-segments / CT-trunk length (cm^2 / cm)	Ile de France	10	91.11	7.33	0.0064 **
	Hung. Merino	10	82.89	2.33	
Fat area of carcass cuts/carcass weight (cm^2/ kg)	Ile de France	10	109.21	18.43	0.1664
	Hung. Merino	10	94.98	25.19	
S/EUROP fat deposition	Ile de France	10	5.70	0.48	0.3923
	Hung. Merino	10	5.30	1.34	
S/EUROP body conformation	Ile de France	10	8.44	0.73	0.0000 ***
	Hung. Merino	10	5.90	0.74	
Roast meat (kg)	Ile de France	10	2.99	0.22	0.0008 ***
	Hung. Merino	10	2.57	0.20	
Rost meat proportion (kg)	Ile de France	10	3.96	0.34	0.0086 **
	Hung. Merino	10	3.56	0.22	
Non-roast meat (kg)	Ile de France	10	2.63	0.31	0.0672
	Hung. Merino	10	2.40	0.21	
Non-roast meat proportion (kg)	Ile de France	10	3.47	0.39	0.0304 *
	Hung. Merino	10	3.10	0.32	
Roast meat parts / non-roast meat parts %	Ile de France	10	114.60	7.72	0.0009 ***
	Hung. Merino	10	111.70	6.15	
Lean meat / right half %	Ile de France	10	54.50	0.46	0.0287 *
	Hung. Merino	10	49.70	0.39	
Pistol leg / right half %	Ile de France	10	43.82	0.02	0.0448 *
	Hung. Merino	10	42.20	0.01	

[1]t-test significance levels: $P \leq 0.05$ (*), $P \leq 0.01$ (**), $P \leq 0.001$ (***)

The butchers' cut

In the quantity of the first class (roast) meat parts and the joint parts the advantage of Ile de France lambs could be statistically confirmed (Table 3). These lambs also showed better results in roast / non-roast parts and lean meat / right carcass ratios comparing to the data of Hungarian Merinos ($P \leq 0.001$). Concerning the most valuable meat cuts (pistol thigh) the advantage of Ile de France lambs was proved as well ($P \leq 0.05$). Among all of the studied parameters, only non-roast meat cuts did not show significant difference between the two breeds.

Conclusions

Having these results the following conclusions could be drawn:

- Our results could confirm that the effect of Ile de France on the slaughter value of Hungarian Merino might be beneficial.
- It is especially true in case of the first class meat cuts.
- It could be finally concluded that CT results could reliably be matched with the results of slaughterhouse measurements.

References

Bajadoux, J., 1954. La race ovine d'Ile de France. Thèse pour le doctorat vétérinaire.

Bouix, J., 1988. Déterminisme génétique des qualités de carcasse des ovines. 3ème Congrès Mondial de Reproduction et de Sélection des Ovins et Bovins, Paris, France, 397-413.

Bouix, J., B. Bibe, C. Lefevre and F. Eychenne, 1986. Variabilité génétique entre et intra-race de la croissance et des qualités de carcasse d'agneau. Journées de la Recherches Ovine et Caprine, Paris, France, 115-145.

Charlet, P. and J. Bougler, 1981. Races ovines. Techniques agricoles.

Berényi, E. and Gy. Kövér, 1991. PC alapú posztprocesszáló program. Pannon Agrártudományi Egyetem [PC based postprocessing software; Pannon University of Agricultural Sciences], Kaposvár, Hungary.

Hounsfield, G.N., 1980. Computed medical imaging. Nobel lecture, Dec. 8, 1979. Journal of Computer Assisted Tomography, Volume 4, 665-674.

Kukovics, S., 2000. Ile de France sheep. In: Jávor, A. and L. Fésüs (eds.), Tenyésztési és fajtahasználati útmutató [Guide for sheep breeding and breed utilisation], Lícium-Art Könyvkiadó és Kereskedelmi Kft, 43-46.

Mezőszentgyörgyi, D., 2000. Különböző genotípusú juhok izom- és faggyúbeépülésének vizsgálata komputeres tomográfia segítségével. Ph.D. értekezés [Study of meat and and fat infiltration into the different genotypes f sheep using computer tomography; PhD Thesis, University of Kaposvár], Kaposvár, Hungary.

Rózsáné Várszegi, Zs., 2003. Keresztezett bárányok vágott test- és húsminősége. Ph.D. értekezés [Meat and carcass quality of crossbred lambs; PhD Thesis; University of Debrecen], Debrecen, Hungary.

Toldi, Gy., 2003. Az S/EUROP minősítés és a juhok vágóértéke közötti összefüggés. Ph.D. értekezés [Relationships between the slaughter value of sheep and the results of S/EUROP classification, PhD Thesis, University of Kaposvár], Kaposvár, Hungary.

Quality classification of the muscle *rectus abdominis* of Charolais Heifers

Marie-Pierre Oury[1], J. Agabriel[2], J. Blanquet[1], D. Micol[2], B. Picard[2], H. Labouré[1], M. Roux[1] and R. Dumont[1]
[1]ENESAD, 26 bd Docteur Petitjean, 21 079 Dijon Cedex, France
[2]INRA, Unité de Recherche sur les Herbivores, Theix, 63 122 St-Genès-Champanelle, France

Abstract

Ninety-nine Charolais heifers were used to study the variability of meat quality attributes in relation to muscle characteristics and heifers slaughter performances. Performances at slaughter (age, carcass weight, development marks, conformation, and fatness) were noted and *rectus abdominis* muscles were excised 24 hours *post mortem*. Meat quality was characterized by sensory evaluations after a 14 days ageing period. A 16-member trained jury scored descriptors of initial and overall tenderness, juiciness and flavour intensity for each meat sample. The physicochemical measurements included shear force, μ-calpain and proteasome sub-unit contents, intramuscular fat content, dry matter content, collagen content, collagen solubility, lactate deshydrogenase, isocitrate deshydrogenase and cytochrome c oxydase activities, mean fibre area, MyHC isoforms, haem iron content and CIELAB colour measurements. A principal component analysis and a hierarchical cluster analysis were used to establish a typology of meat depending on their sensorial attributes. Five meat types (I, II, III, IV and V) were classified by increasing tenderness from the type I to the type V. The types IV and V led to both tender and juicy meat and contained the most intramuscular fat (20% DM). The lowest level of tenderness in some meat samples (type II) may in part be explained by a high shear force linked to poor solubility and a high quantity of collagen. Meat from types I and II stemmed from the oldest heifers (35.0 to 34.5 months). The increase of average tenderness between the five meat types is parallel to that from the daily weight gain of the animals during their life (625 to 682 g/day of life).

Keywords: quality, meat, muscle type

Introduction

A better control of meat quality has a major importance for beef retailers. Indeed, the insurance of a certain meat quality and the decrease of its variability could permit to satisfy the consumer's requirement for a consistently satisfactory product (Bindon and Jones, 2001). Meat quality is commonly evaluated by the descriptors of tenderness, juiciness and flavour intensity. Most of the bibliography indicated a positive correlation between these quality traits in the muscle *longissimus thoracis*. For example, between tenderness and juiciness, a correlation varying between +0.54 and +0.58 was found. Between tenderness and flavour intensity, the correlation varied between +0.29 and +0.34. Moreover, juiciness and flavour intensity were positively correlated with a coefficient varying between +0.20 and +0.70 (Gregory *et al.*, 1994; Dransfield *et al.*, 2002; Rhee *et al.*, 2004; Serra *et al.*, 2004). Thus, the correlations of +0.35 to +0.50 between meat quality traits explained that different levels of juiciness and flavour intensity might be found within one level of tenderness.

The assessment of meat quality by sensory evaluations is expensive. A need exists also to find slaughter performances and / or muscular physicochemical characteristics related to sensory attributes that are easy to measure in order to select directly the carcasses.

The first aim of the present study was to quantify the variability in meat quality traits, by constituting a typology of meat by multifactorial statistical analysis on tenderness, juiciness and flavour intensity. Then, the second aim was to explain the typology both by slaughter performances of heifers and physicochemical characteristics of muscles.

Material and methods

Animals, slaughter performances and muscle samples

This study was conducted on animals of the same breed and gender slaughtered under the same conditions. A total of 99 Charolais heifers were collected in a cooperative of livestock farmers located in Venarey-Les-Laumes (Côte-d'Or, France). All slaughters occurred in the same industrial slaughterhouse in order to standardize slaughtering, chilling and storing procedures.

The day before slaughter fatness scores were noted by alive scoring. The average live weight gain was calculated using the slaughter age and the cold carcass weight. The animals were slaughtered between 26 and 43 months of age with an average age of 33.4 months. Carcasses weighted on average 381 kg (330 – 509 kg) and were scored 3 on 5 for fatness.

The pH of each carcass was measured at 24 h *post mortem* in the *rectus abdominis* (RA) muscles. The colour of RA muscle was measured 24 hours *post mortem* using a portable spectrophotometer (CR300, MINOLTA). Colour coordinates were calculated in the CIELAB system and results were expressed as lightness (L*), redness (a*) and yellowness (b*).

Physicochemical characteristics and sensory evaluation

Samples of RA muscle were removed from the carcasses 24 hours after slaughter for sensory evaluation and physicochemical measurements.

Dry matter content was measured by oven drying at 103 °C for 48 hours. The intramuscular lipid content was determined by the Soxhlet standard method (NF V 04-402, 1968). The haem iron content and the collagen content were evaluated respectively by the method of Hornsey (1956) and by measurement of hydroxyproline content (collagen = 7.5 x hydroxyproline) according to the method of Bergman and Loxley (1963). Collagen in the insoluble part was determined according to a procedure given by Bonnet and Kopp (1992). Collagen solubility was expressed as the percentage of heat-soluble collagen (total – insoluble) to total collagen.

The metabolic muscle type was determined by measuring enzyme activities. The anaerobic glycolytic metabolism was assessed by lactate dehydrogenase (LDH) activity (Ansay, 1974). The aerobic oxidative metabolism was assessed by isocitrate dehydrogenase (ICDH) and cytochrome-c oxydase (COX) activities according to the method of Briand *et al.* (1981) and Piot *et al.* (1998).

Muscle mean fibre area was determined on 10-μm thick sections cut perpendicular to the muscle fibres with a cryotome at –25 °C. Between 100 and 200 fibres in each of two different locations in the muscle were used to determine the mean fibre area by computerized image-analysis. The different types of myosin heavy chains isoforms were determined on the basis of previously determined migration pattern (Young and Davey, 1981; Picard *et al.*, 1995) on sodium Dodecyl Sulfate polyacrylamide gel electrophoresis (SDS-PAGE) performed by the method of Laemmli (1970).

Contents of μ-calpains and proteasome 27K sub-unit were obtained by western-blot. The protein concentration of the sample was determined according to the method of Bradford (1976) using bovine serum albumin as standard. The separation gel was a 12.5% polyacrylamide gel (C = 2.6%) (Towbin *et al.*, 1979). Migration was performed at 4 °C at 150 V. The proteins were also electro transferred from the SDS gel onto a membrane of polyvinyl. The membrane was saturated 1 hour. Incubation with the anti μ-calpain and anti 27K antibodies, diluted in a solution containing TBS 1X and 10% Tween was performed during one hour at 37 °C under shaking. The dilution varied depending on antibody used (1 / 8,000 or 1 / 4,000). The second antibody (anti mouse igG) was diluted 10,000 fold and applied for 1 hour at room temperature. After washing in TBS 1X Tween 10% and using luminal reagent, development was made on a photo film and analysed with Quantitione® software (Biorad).

The textural properties of raw meat were studied in the RA muscle, vacuum packaged and aged at 4 °C for 14 days *post mortem*. Shear force was measured according to the method of Salé (1971).

Sensory assessment was conducted on samples of RA muscle by a test panel. The aim was to classify the 99 heifers by a monadic test. The RA muscles were vacuum-packaged and chilled for 14 days at +4 °C for ageing. After ageing, the RA muscle was trimmed and cut into 1.5 cm thick homogeneous steaks, then vacuum-packaged a second time and frozen at –20 °C until the sensory analysis. Meat in the form of 15 mm steaks was thawed and then cooked for 1 minute 45 seconds in an Infra grill Duo Sofraca set at a temperature of 300 °C. After cooking, the steaks were cut into 20 mm samples that were served at an internal temperature of 55 °C. The jury trained in meat appreciation consisted of 16 panellists. The panellists evaluated the cooked samples for initial tenderness, overall tenderness, juiciness and flavour intensity. Initial tenderness was defined, as tenderness felt at the first bite, whereas overall tenderness indicated the amount of chewing needed before swallowing a meat sample. Each attribute was rated on a 10-point non-graduated scale on an ascending scale of quality with a score of 0 equivalent to though, lacking juiciness and low flavour, and a score of 10 equivalent to tender, extremely juicy and high flavour. Therefore, at each session, a monadic presentation of 5 samples was done. The sessions were carried out in a sensorial analysis room with individual boxes, under artificial non-coloured lighting.

Statistical analyses

Principal Component Analysis (PCA) implemented in the WINSTAT software was used (ITCF-CIRAD, 1997). Afterwards, the first three factors obtained in the PCA were included in a Hierarchical Cluster Analysis (HCA, Winstat). The item coordinates on the first three dimensions were also used in the HCA. Indeed, it is quite difficult to make visual items or variable categories on more than two factorial dimensions. The centroid method was used as the means of aggregation. The programme identified the cluster, which has the minor within-group variance and the greater variance between groups. Five meat types were identified using their description by the three factorial dimensions of the PCA.

Among the five types, heifers slaughter performances and muscular characteristics differences were evaluated. Variance analysis and mean multiple comparisons with one factor were carried out using the General Linear Model procedure (LSMEAN – PDIFF) from SAS 9.1 (SAS Institute, 2002).

Results

Initial and overall tenderness scores were distributed between 3.0 and 7.8 /10. The average scores were respectively equal to 5.7/10 and 5.3/10. The lower score for overall tenderness than for initial tenderness indicated that *rectus abdominis* meat samples used to be tender at the first chewing and then a tougher after many chewing. Juiciness scores were also distributed between 3.0 and 7.8 /10 with an average score of 5.3/10. Flavour intensity was scored between 4.3 and 6.4 /10 with an average of 5.7/10. Thus, the variability of tenderness and juiciness descriptors was twice higher than those from flavour intensity (CV = 7.0 vs. 14.0 to 17.0%). As expected, initial and overall tenderness were highly correlated (+0.97). Juiciness was positively correlated with initial and overall tenderness (+0.52 and +0.46) and with flavour intensity (+0.48). Flavour intensity was also positively correlated with initial and overall tenderness (+0.40 and 0.35) (Table 1).

A principal component analysis (PCA) was carried out on the scores of the four quality traits. The first three significant dimensions explained 99.0% of the total inertia, with respectively 75.6% for the first dimension, 18.6% for the second one and 4.8% for the third one. The PCA compared on the first dimension high initial and overall tenderness (negative coordinates) to low initial and overall tenderness (positive coordinates). Juiciness and flavour intensity descriptors were well positioned on the second and third dimensions respectively, with increasing scores when going near the negative coordinates (Figure 1). Nevertheless, the conclusions about the third dimension had to be moderated, as this dimension explained only a small part of total inertia (4.8%).

A hierarchical cluster analysis (HCA) allowed to organise samples of meat in types, depending on their sensorial quality. The classification was done on the samples coordinates on the first three significant dimensions. The PCA and the HCA led to a typology of meat divided into five types classified from the type I (low tenderness) to the type V (high tenderness) by increasing tenderness. Meat from type I had both a rather low tenderness and a low juiciness but their flavour intensity was intermediate. Those from type II had a rather low tenderness and a medium juiciness but their flavour intensity was high. Meat from type III had both low tenderness, juiciness and flavour intensity scores. Those from type IV were characterised by a medium tenderness, a high juiciness and a medium flavour intensity. To finish with, meat from type III were described as rather tender, with a medium juiciness and a high flavour intensity (Table 2). As expected, the first dimension opposed

Table 1. Correlation coefficients between meat quality attributes of m. rectus abdominis (n=99).

	Initial tenderness	Overall tenderness	Juiciness
Overall tenderness	+0.97*		
Juiciness	+0.52*	+0.46*	
Flavour intensity	+0.39*	+0.34*	+0.47*

*Correlations are significant (p>0.05).

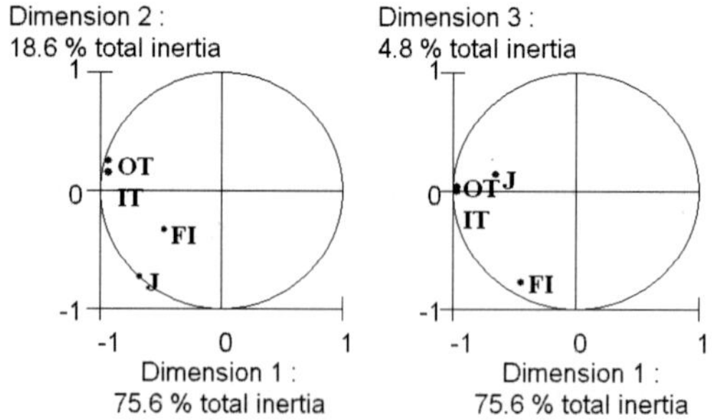

Figure 1. Representation of meat quality attributes in the two-dimensional space of the principal component analysis defined by factors 1 and 2 and factors 1 and 3 (IT: initial tenderness; OT: overall tenderness; J: juiciness; FI: flavour intensity).

Table 2. Meat quality attributes, muscle physicochemical characteristics and heifers performances at slaughter depending on meat types.

	Mean	Meat types					P
		I	II	III	IV	V	
N	99	26	12	16	20	25	
Meat quality attributes							
Initial tenderness (score on 10)	5.7	4.97^{a}	5.10^{ab}	5.48^{b}	5.92^{c}	6.55^{d}	0.001
Overall tenderness (score on 10)	5.3	4.54^{a}	4.52^{a}	5.14^{b}	5.54^{c}	6.26^{d}	0.001
Juiciness (score on 10)	5.3	4.51^{a}	5.86^{b}	4.78^{a}	5.94^{c}	5.58^{b}	0.001
Flavour intensity (score on 10)	5.7	5.72^{b}	5.97^{c}	5.16^{a}	5.64^{b}	6.05^{c}	0.001
Physicochemical characteristics							
Shear force - raw meat (daN)	6.7	6.7^{a}	7.9^{b}	6.3^{a}	6.5^{a}	6.5^{a}	0.038
Total collagen (mg/g dry matter)	21.3	21.0^{a}	23.9^{b}	22.0^{a}	20.9^{a}	20.1^{a}	0.011
Soluble collagen (% total collagen)	16.1	16.7^{b}	13.8^{a}	16.6^{b}	16.5^{b}	15.5^{ab}	0.053
µ-calpain (% of a control)	100	97	94	99	99	109	0.374
27K proteasome sub-unit (% of a control)	145	152	132	156	145	138	0.434
Intramuscular fat (% dry matter)	17.8	15.9^{a}	17.0^{ab}	15.0^{a}	20.3^{b}	19.9^{b}	0.007
Mean fibre area (µm²)	3380	3600	3383	2805	3314	3568	0.110
Myosin heavy chain I (%)	30.2	29.1	30.3	29.5	30.5	30.8	0.689
Myosin heavy chain IIa (%)	35.2	34.9	36.2	35.7	35.4	36.3	0.955
Myosin heavy chain IIx (%)	34.6	35.9	33.5	34.8	34.1	32.9	0.600
Isocitrate dehydrogenase (µmol/min/g)	1.39	1.41	1.53	1.43	1.31	1.33	0.407
Lactate dehydrogenase(µmol/min/g)	501	498	498	501	506	503	0.994
Cytochrome c oxydase (µmol/min/g)	16.6	16.6	15.7	17.4	15.8	17.2	0.565
Haem iron content (µg/g dry matter)	59.2	61.0	61.5	59.4	57.5	58.5	0.500
L*	33.7	34.0	32.3	34.6	33.1	33.7	0.221
a*	20.3	20.1	21.1	20.1	20.1	20.4	0.423
b*	5.8	5.7	5.5	5.8	5.8	5.8	0.991
Slaughter performances							
Slaughter age (month)	33.4	35.0^{b}	34.5^{ab}	32.2^{a}	32.9^{ab}	32.2^{a}	0.081
Carcass weight (kg)	381	382^{ab}	378^{ab}	366^{a}	395^{b}	382^{ab}	0,096
Average live weight gain (g/d of life)	664	625^{a}	638^{ab}	641^{ab}	688^{b}	682^{b}	0,038
Fatness score – on alive animals – (on 10)	6,9	6,73	6,67	6,88	6,85	7,08	0,397

Within row, means with an identical superscript are not significantly different (p>0.05).

types I (positive coordinates) and V (negative coordinates). On this dimension, types II, III and IV were located successively between types I and V. Type IV, characterized by high juiciness scores was located on the negative coordinates of the second dimension. On the contrary, types I and III were located on the positive coordinates of the second dimension, as these types were characterised by low juiciness scores. Type III was characterised by low flavour intensity scores and was also located on the positive coordinates of the third dimension, whereas types II and V were located on the negative coordinates of the third dimension.

Physicochemical characteristics of each meat type were compared in order to establish those that could explain the differences. The studied physicochemical characteristics were not sufficient to characterise each meat type, but allowed nevertheless to differentiate some types from the others (Table 2). Criteria that discriminated meat types were dry matter content, intramuscular fat content, shear force and collagen content.

Meat from types IV and V, having highest scores for each descriptor than those from types I and III, had highest intramuscular fat (20.3 and 19.9 vs. 15.9 to 17.8% DM) contents. Thus, the intramuscular fat content seems to be the factor that discriminate the most the meat types. Meat from type II had the highest content in total collagen in comparison to the others types (23.9 vs. 20.1 to 22.0 mg/g DM). In addition, these meats have the lowest content in soluble collagen (13.8 vs. 15.5 to 16.7% total collagen). Type II led the highest shear forces in comparison to the others types (7.9 vs. 6.3 to 6.5 daN). Nevertheless, meats from type II have the same tenderness scores than meats from types I and III. Indeed, the highest intramuscular fat content of meat from type II in comparison to the types I and III (17.0% DM vs. 15.9% and 15.0% DM) could have had a positive impact on tenderness scores. The five meat types had similar characteristics for mean fibre area, myosin heavy chain I, IIa and IIx percentages, isocitrate dehydrogenase, lactate dehydrogenase and cytochrome c oxydase activities, μ-calpain and 27K proteasome sub-unit content, haem iron content and colour measurements (L*, a*, b*).

The five meat types came from heifers that have significantly different slaughter ages, carcass weights and average live weight gains (Table 3). Meat from type I came from the oldest heifers (35.0 months) whereas meat from types III and V came from the youngest ones (32.2 months). Heifers that gave meat from type III were slaughtered at the lightest carcass weight (366 kg) whereas those that gave meat from type IV were slaughtered at the heaviest carcass weight (395 kg). Heifers that gave meat from types I, II and V have an intermediate carcass weight. The tenderness gradient between types I and V was in parallel to this from heifers average live weight gain, going from 625 g/d of life for heifers from type I to 382 g/d for heifers from type V (Table 3). Fatness and the muscular development scores of heifers were equal among the five meat types ($p=0.397$ and $p=0.315$).

Discussion and conclusion

The variability of the tenderness and juiciness scores was twice higher than those from the flavour intensity ones as already observed on the muscle *longissimus thoracis* (LT) (Renand *et al.*, 1997, 2001; Rhee *et al.*, 2004). Thus, as meat quality attributes of tenderness, juiciness and flavour intensity were positively correlated, juiciness and flavour intensity scores have probably received a positive influence of the favourable assessment of tenderness scores. Nevertheless, correlations being between +0.34 and +0.52, different levels of juiciness and flavour intensity may be obtained among one level of tenderness. Indeed, within the same level of tenderness, meats from type I were characterized by low juiciness and medium flavour intensity whereas meats from type II were characterized by medium juiciness and high flavour intensity. Moreover, within the same level of tenderness, meats from type III were characterized by low juiciness and flavour intensity whereas meats from type II were characterized by medium juiciness and high flavour intensity.

The intramuscular fat content firstly distinguished the five meat types. Indeed, types IV and V, characterized by high tenderness and medium to high juiciness and flavour intensity scores, had higher fat content than types I, II and III. This observation may be explained by the positive correlation that appears in the LT muscle between intramuscular fat content and meat quality traits of tenderness and juiciness (Silva *et al.*, 1999; Renand *et al.*, 2001; Dransfield *et al.*, 2002; Geay *et al.*, 2002; Jeremiah *et al.*, 2003).

Secondly, meat types were distinguished by shear force values. Indeed meats from type II had a significantly higher shear force than those from type I, III, IV and V. As shear force is known to be respectively positively and negatively correlated with total and soluble collagen content (Silva *et al.*, 1999; Renand *et al.*, 2001; Rhee *et al.*, 2004), type II was characterized by higher total collagen and lower soluble collagen contents than types I, III, IV and V. According to the literature, shear force and total collagen content were negatively correlated with tenderness, whereas soluble collagen content was positively correlated with this descriptor (Whipple *et al.*, 1990; Gregory *et al.*, 1994; Silva *et al.*, 1999; Renand *et al.*, 2001; Jeremiah *et al.*, 2003; Rhee *et al.*, 2004).

Of course meat from type II was characterized by a lower tenderness in comparison to meat from types IV and V, but type II led to equivalent tenderness scores than types I and III. The intermediate level of intramuscular fat of meat from type II may have offset a part of the negative impact of shear force and collagen content. This hypothesis is improved by the highest juiciness and flavour intensity scores for meat from type II than those from types I and III.

No differences were found between meat types on proteases content. In the literature, some papers indicated no parallel evolution between proteases activities and tenderness scores (Whipple *et al.*, 1990; Renand *et al.*, 2001; Serra *et al.*, 2004).

The metabolic and contractile properties of muscles were not significantly different among the five meat types. Fibres types are significantly correlated nether with tenderness nor with juiciness (Whipple *et al.*, 1990; Renand *et al.*, 2001; Serra *et al.*, 2004). Nevertheless, when considering the metabolic enzymes, tenderness seems to be negatively correlated with ICDH activity (Renand *et al.*, 2001). This conclusion was also found in this work, with a lower activity for types IV and V in comparison to those from types I, II and II. The higher oxidative metabolism of muscles from types IV and V is logical with the higher content in intramuscular fat, principal source of the cell energy.

Meat types were differentiated in part by physiochemical characteristics and especially by muscles content of intramuscular lipids, total collagen and soluble collagen and shear force. The low differentiation of meat types by physicochemical characteristics may be linked to the low explanation level of each quality descriptors by the physicochemical characteristics. This is almost right when considering an homogeneous population of heifers and a muscle like the *rectus abdominis*, considered as a muscle with many specificities, especially on muscular fibres, that could have reduced the differences between meat types (Picard *et al.*, 2003). Moreover, according to Hocquette *et al.* (2004), characteristics that explain tenderness may not be the same than those explaining juiciness and/or flavour intensity.

For a given age, heifers with the highest carcass weight gave the tenderest meat but also the highest scores in juiciness and flavour intensity. Indeed, increasing the carcass weight was in favour of meat quality. Sanudo *et al.* (2004) have already found this conclusion when comparing entire males slaughtered between 300 and 550 kg. These authors found that increasing slaughter weight led to a significant decrease of Warner-Bratzler shear force and meat toughness. For a given carcass weight, the youngest heifers led to the most tender meat. The negative correlation between slaughter age and meat tenderness was also confirmed in this study (Renand *et al.*, 2001). Indeed, the tenderest meat came from heifers with an average live weight gain higher than 670 g/d. The less tender ones came from heifers with an average live weight gain lower than 640 g/d. In the studied population, there were no heifer slaughtered both old (37 months) and with a high carcass weight (higher than 450 kg). This kind of animal would have allowed to test the respective effects of slaughter age and carcass weight on the tenderness determinism. Concerning juiciness scores, when intramuscular fat content of meat was low (less than 16.0% DM), the sensorial jury could not distinguish the juiciness differences. As intramuscular fat leads to a decrease of cooking losses and reduce the sensation of dry meat (Renand *et al.*, 2001), juiciness scoring of meat containing low fat, could have been affected. However, when the intramuscular fat content was higher (17.0 to 20.3% DM), the jury was able to distinguish different levels of juiciness. Thus, juiciness scores increased in parallel to carcass weight, whatever slaughter age. This result needs a confirmation, as juiciness is a sensorial descriptor less studied than tenderness or flavour intensity (Geay *et al.*, 2002). Concerning flavour intensity descriptor, conclusions about animals slaughtering performances were difficult to establish because of the low variability of flavour intensity scores. Nevertheless, as sensorial descriptors are positively correlated the ones with the others, carcass weights were found positively correlated with flavours intensity, as already found by Renand *et al.* (2001).

References

Ansay, M., 1974. Individualité musculaire chez le bovin. I. Enzymes. Annales de Biologie Animale Biochimie Biophysique, Volume 14, 471-486.

Bergman, I. and R. Loxley, 1963. Two improved and simplified methods for the spectrophotometric determination of hydroxyproline. Analytical Chemistry, Volume 35, 1961-1965.

Bindon, B.M. and N.M. Jones, 2001. Cattle supply, production systems and markets for Australian beef. Australian Journal of Experimental Agriculture, Volume 41, 861-877.

Bonnet, M. and J. Kopp, 1992. Préparation des échantillons pour le dosage et la caractérisation qualitative du collagène musculaire. Viandes et Productions Carnés, Volume 13, 87-91.

Bradford, M., 1976. A rapid and sensitive method for the quantitation of microgram quantities of protein utilizing the principle of protein-dye binding. Analytical Biochemistry, Volume 72, 248-254.

Briand, M., A. Tamant, Y. Briand, G. Monin, B. Durand, 1981. Metabolic types of muscle in the sheep. I. Myosin ATPase, glycolytic and mitochondrial enzyme activities. European Journal of Applied Physiology, Volume 46, 347-358.

Dransfield, E., A. Abouelkaram, D. Bauchart, J. Culioli, C. Jurie, J. Lepetit, A. Listrat, J.F. Martin and B. Picard, 2002. Qualité sensorielle de la viande provenant de 3 muscles de taurillons et de vaches de réforme de 4 races allaitantes du Massif Central. Rencontres Recherches Ruminants, Volume 9, 255-258.

Geay, Y., D. Bauchart, J.F. Hocquette and J. Culioli, 2002. Valeur diététique et qualités sensorielles des viandes de ruminants - Incidence de l'alimentation des animaux INRA Productions Animales, Volume 15, 37-52.

Gregory, K.E., L.V. Cundiff, R.M. Koch, M.E. Dikeman and M. Koohmaraie, 1994. Breed effects, retained heterosis, and estimates of genetic and phenotypic parameters for carcass and meat traits of beef cattle Journal of Animal Science, Volume 72, 1174-1183.

Hocquette, J.F., I. Ortigues-Marty, B. Picard, M. Doreau, D. Bauchart and D. Micol, 2004. La viande des ruminants: de nouvelles approches pour améliorer et maîtriser la qualité. Viandes et Produits Carnés, Volume 24, 7-18.

Hornsey, H.C., 1956. The colour of cooked cured pork. I. Estimation of the nitric oxide-haem pigments. Journal of the Science of Food and Agriculture, Volume 7, 534-540.

ITCF-CIRAD, 1979. WINSTAT software, Version 2, Paris, France.

Jeremiah, L.E., M.E.R. Dugan, J.L. Aalhus and L.L.Gibson, 2003. Assessment of the relationship between chemical components and palatability of major beef muscles and muscle groups. Meat Science, Volume 65, 1013-1019.

Laemmli, U.K., 1970. Cleavage of structural proteins during the assembly of the head of bacteriophage T4. Nature, Volume 227, 680-685.

Picard, B., J. Robelin and Y. Geay, 1995. Influence of castration and postnatal energy restriction on the contractile and metabolic characteristics of bovine muscle. Annales de Zootechnie, Volume 44, 347-357.

Picard, B., M.P. Oury, C. Jurie, I. Cassar-Malek and J.F. Hocquette, 2003. Variabilité inter-musculaire des propriétés contractiles et métaboliques de muscles de bovins. Premier Congrès de la Société Française de Nutrition, Clermont-Ferrand. Cahier de Nutrition et Diététique 2004, Volume 39, 70.

Piot, C., J.H. Veerkamp, D. Bauchart, J.F. Hocquette, 1998. Contribution of mitochondria and peroxisomes to palmitate oxidation in rat and bovine tissues. Comparative Biochemistry and Physiology - Part B Biochemistry and Molecular Biology, Volume 121, 185-194.

Renand, G., C. Touraille, Y. Geay, P. Berge, J. Lepetit and B. Picard, 1997. Variabilité des qualités organoleptiques de la viande bovine en relation avec les caractéristiques musculaires. Rencontres Recherches Ruminants, Volume 4, 311-314.

Renand, G., B. Picard, C. Touraille, P. Berge and J. Lepetit, 2001. Relationships between muscle characteristics and meat quality traits of young Charolais bulls. Meat Science, Volume 59, 49-60.

Rhee, M.S., T.L. Wheeler, S.D. Shackelford and M. Koohmaraie, 2004. Variation in palatability and biochemical traits within and among eleven beef muscles Journal of Animal Science, Volume 82, 534-550.

SAS, 2002. SAS Institute Inc., Cary, NC, USA.

Salé, P., 1971. Bulletin Technique C.R.Z.V. Theix, INRA, Volume 6, 35-44.

Sanudo, C., E.S. Macie, J.L. Olleta, M. Villarroel, B. Panca and P. Alberti, 2004. The effects of slaughter weight, breed type and ageing time on beef meat quality using two different texture devices. Meat Science, Volume 66, 925-932.

Serra, X., M. Gil, M. Gispert, L. Guerrero, M.A. Oliver, C. Sanudo, M.M. Campo, B. Panea, J.L. Olleta, R. Quintanilla and J. Piedrafita, 2004. Characterisation of young bulls of the Bruna dels Pirineus cattle breed (selected from old Brown Swiss) in relation to carcass, meat quality and biochemical traits. Meat Science, Volume 66, 425-436.

Silva, J.A., L. Patarata and C. Martins, 1999. Influence of ultimate pH on bovine meat tenderness during ageing. Meat Science, Volume 52, 453-459.

Towbin, H., T. Staehelin and J. Gordon, 1979. Electrophoretic transfer of proteins from polyacrylamide gels to nitrocellulose sheets: procedure and some applications. Proceedings of the National Academy of Sciences of the United States of America, Volume 76, 4350-4354.

Whipple, G., M. Koohmaraie, M.E. Dikeman and J.D. Crouse, 1990. Predicting beef-longissimus tenderness from various biochemical and histological muscle traits. Journal of Animal Science, Volume 68, 4193-4199.

Young, O.A. and C.L. Davey, 1981. Electrophoretic analysis of proteins from. single bovine muscle fibres. Biochemical Journal, Volume 195, 317-327.

Ethanol specimen of beef muscle samples for NIRS is rapid but as efficient as the freeze-dried sample preparations

Giorgio Masoero, M. Iacurto and G. Sala
C.R.A. Istituto Sperimentale per la Zootecnia, Via Pianezza 115, 10151 Torino, Italy

Abstract

Rapid muscle preparation and conservation for NIRS in ethanol has been studied. In a factorial 3x3x2 trial (N=36) two homogeneous sub-apical *Semitendinosus* bovine muscle steaks were sliced, (20mm*Ø20mm) and plugged in a number of 1-2-3, into a 50ml tube containing 30 ml of ethanol 95.0, 63.3 or 31.7%, after exposition, or no exposition to an O_3 flux. At 0h and 24h, the samples were weighed, tested by a penetrometer and then scanned by NIRS. The low ethanol concentration gave both a lower weight (0.24) and a lower tenderness loss (0.23) while the two higher concentrations gave a higher and homogeneous weight (0.35 and 0.34) and tenderness loss (0.64 and 0.59), the two traits being quadratically correlated (R^2=0.67). The O_3 treatment was not perceived by NIRS nor by the variables, while the three ethanol concentrations were clearly distinguished (R^2_{val}=0.94) by NIRS. The number of specimens in one tube was related to NIRS (R^2_{val}=0.26) but not to the variables. Thus it could be recommend to place a maximum of 15g/30ml 95% ethanol into a tube, whether in single or multiple specimen, even from different tissues, but of similar size. An application of this part of the discrimination trial was carried out with muscle specimens, obtained from 54 Piemontese and 16 Italian Friesian cattle, which were immersed in ethanol or were freeze-dried. In *Semitendinosus* m., the classification ability of the NIRS for the two breeds reached 97% in ethanol *vs*. 94% in freeze-dried preparations. In the *Sternomandibularis* m. ethanol preparation, the breed identification was 91%. It can be concluded that this new simple and rapid method, very efficacious in separating systematic factors, offers also a good level of efficiency. Further studies are necessary to link this specimen to trait and treatment effects.

Keywords: NIRS, muscle, ethanol, cattle breeds, intrinsic characteristics

Introduction

Meat researchers have long sought non-destructive, objective techniques to predict meat quality. Several studies have stated that near-infrared reflectance spectroscopy (NIRS) can be used to predict beef tenderness. Shackelford *et al.* (2004), and also Cozzolino and Murray (2004), capitalized on the advances in instrumentation, that now allow a single instrument to scan the entire electromagnetic spectrum (350–2500 nm, UV-Vis-NIRS), developing a repeatable, non-destructive technique for spectroscopic evaluation of meat quality.

Several Authors have investigated off-line experimental procedures which are destructive in that they require excision of a muscle sample for spectroscopy (Hildrum *et al.*, 1994). Our work is more concerned with methods for off-line or at-line measurements, thus we are interested in obtaining easy and stable samples of specimens from animal muscle. Ethanol preparation of specimens for NIRS was first utilized in rabbits (Masoero *et al.*, 2004), and then was replicated in buffalo (Masoero *et al.*, 2005) allowing significant, easy and rapid NIRS discrimination results of the systematic main factors to be obtained.

Material and methods

In a factorial 3x3x2 trial (N=36) two homogeneous sub-apical *Semitendinosus* bovine muscle steaks were sliced, (20mm*Ø20mm, average 4.9±0.7g) and plugged, in a number of 1,2 or 3 into a 50ml tube containing 30 ml of ethanol 95.0, 63.3 or 31.7%, after exposition, or no exposition, of the steaks for 5 min to an O_3 flux (20ppm). At 0h and 24h, the specimens were weighed and contextually tested by a penetrometer (Schiltknecht, Ing., Zurich) with repeated needle perforation (n=10) in each tube. After the ethanol (ETH) treatment, one specimen was then scanned, for each tube, by using a FT-NIRS device (Quantum-One, PE), from 1000 to 2500 nm, that is, 3001 points in the interferograms. All the points were chemometrised by the Modified Partial Least Squares (MPLS) method using WINISI-II from Infrasoft International (ISI, State College, PA, USA), utilising a cross-validation system to assess the optimal number of latent variables to be included into the equations and permitting one passage for elimination of the outlier samples. The main factors, ETH concentration, number of specimens per tube and O3 treatment as well as the variables weight loss (WL), penetrometer score (PEN) and tenderness loss (TL) were calibrated.

In a discrimination trial, samples of two muscles from 54 Piemontese and 16 Italian Friesian young bulls were immersed in ethanol or freeze-dried. The classification ability in the pertinent groups of the NIR spectra was established by using the Discrim tool of the WINISI-II software and also from the binary contrasts on dummy variables.

Results and discussion

The immersion in ethanol abruptly changed the aspect of the muscle, which quickly resembled a 'cooked' specimen. The weight losses after 24 h were on avg. 0.30±0.06% (Table 1); they were minimum at the lower ETH concentration (0.24) and invariant at the two higher concentrations (0.34 and 0.35, respectively) showing a parabolic trend with R^2=0.73 (Figure 1). In parallel to the increased concentration of ETH, the loss of tenderness reached an avg. highly variable level of 0.50±0.24, which was unbalanced between the lower (0.23) and the higher concentrations (0.34 and 0.35) with a significant parabolic trend (R^2=0.68; Figure 2). The parabolic patterns were also evident in the regression of the tenderness loss on the weight loss (R^2=0.67; Figure 3) and on the penetrometer score of the ETH specimens (R^2=0.84; Figure 4). This latter relationship was higher because it was corrected for the raw tenderness proper to each sample or sub-sample.

Table 1. ANOVA of the variables and NIR Spectroscopy of the main effects and of the variables.

Variables	Mean	Std.	Univariate main effects model R^2				NIR Spectroscopy of variables, *R2-val*
			Concentration	Number of specimens	Ozone treatment	Full model with interactions	
Weight loss 24 h	0.30	0.06	0.72	0.04	0.00	0.77	0.68
Penetrometer raw, mm	59.9	9.2					0.11
Penetrometer ethanol, mm	31.9	15.9	0.68	0.06	0.03	0.81	0.82
Tenderness loss	0.50	0.24	0.64	0.03	0.00	0.74	0.66
NIR Spectroscopy	*R2-cal*	1.00	0.32	0.13			
of main effects	*R2-val*	0.94	0.26	0.00			

R^2cal= $R^2_{calibration}$; R^2val= $R^2_{cross\text{-}validation}$.

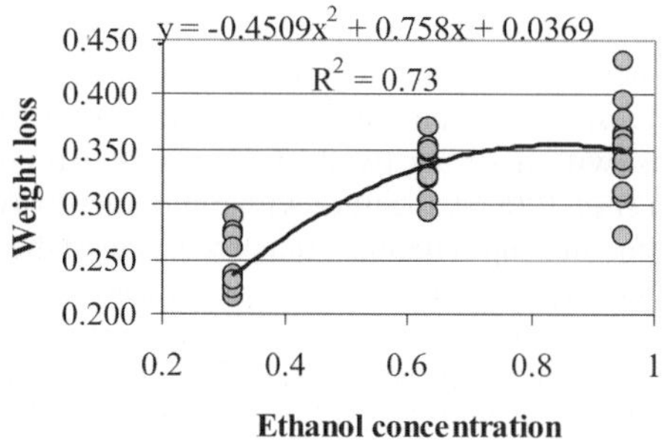

Figure 1. Regression of WL on ETH%.

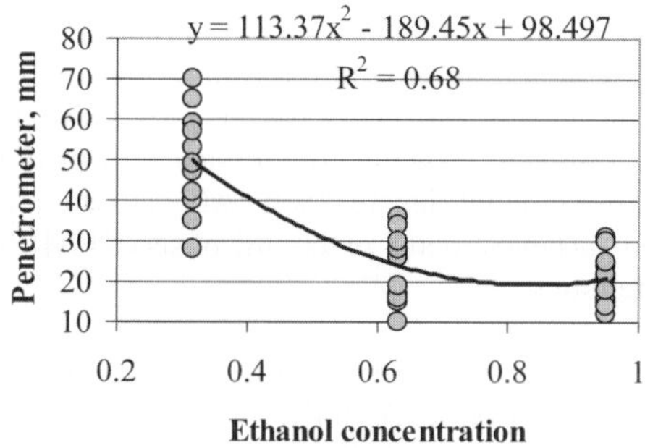

Figure 2. Regression of the PEN scores on ETH%.

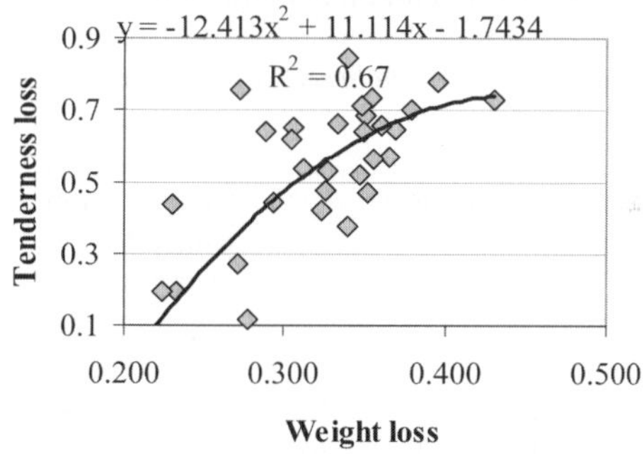

Figure 3. Regression of TL on WL%.

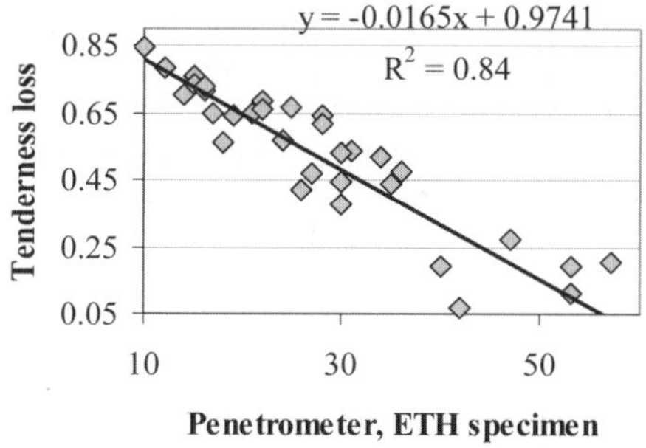

Figure 4. Regression of the TL scores on PEN score.

The absolute high variability observed in the hardness increase suggests two possible explanations: the first due to inherent differences in the apical region of the *Semitendinosus* two-cm width steak; the second due to an individual response of the specimens to ethanol denaturation effects, according to liquid/mass ratio variations, with a strong twist of the fibres. An ethanol-based fixative has recently been used by Iesurum *et al.* (2006), together with rapid microwave-stimulated processing, in *post-mortem* material, resulting in a rapid fixation and processing of the tissues with morphology, histochemical stains, and immunocytochemistry comparable to formalin-fixed material. Furthermore, this alternative fixation gives better DNA recovery in higher amounts compared with DNA extracted from formalin-fixed tissue, which is particularly advantageous in forensic pathology.

The effect of the ETH concentration was better perceived by NIR Spectroscopy than by the measured variables (R^2_{val}=0.94 vs. 0.64-0.72; Table 1). The Ozone treatment appeared, overall, to not have been effective. On the contrary, the number of specimen (1-3) per tube did not affect the weight loss or the tenderness score or loss, but the NIR Spectroscopy appraised this main effect at a not high, but relevant, level (R^2_{val}=0.26). When looking at the calibration of the variables, these appeared to be closely based on the NIR spectra; especially for the PEN score, the R^2_{val}=0.82 assumed by the NIRS was a value that is equivalent to the whole influence of the weight loss on the hardness increase in the muscle mass after ethanol denaturation had occurred. However, because the NIRS did not explain l the measured variables to any extent the observed NIRS effects include other, not externalised, intrinsic characteristics.

In *Semitendinosus* m. the classification ability of the NIRS for the two breeds reached 97% in ethanol vs. 94% in freeze-dried preparations (Table 2). In the *Sternomandibularis* m. ethanol preparation, the breed identification was 91%. These quite optimistic values were attained by using a not so high discriminative equation (R^2_{val} = 0.63; 0.62 and 0.51 respectively), but where also confirmed by the cross-validation process of the MPLS standard binary equations (R^2_{val} = 0.83; 0.82 and 0.79).

Table 2. NIRS classification ability between the breeds of freeze-dried or ethanol prepared muscle specimens.

Specimens	Breed	P	I	Classification ability %		*EQA*	R^2cal	R^2val
					Mean			
Semitendinosus – Freeze-Dried	P	53	1	98	94	D	0.82	0.63
	I	3	13	81		B	0.94	0.83
Semitendinosus – Ethanol	P	53	1	98	97	D	0.96	0.62
	I	1	15	94		B	0.97	0.82
Sternomandibularis – Ethanol	P	53	1	98	91	D	0.55	0.51
	I	5	11	69		B	0.93	0.79

EQA= equation; P = Piemontese; I = It. Friesian; D = MPLS discriminant equation; B = MPLS binary equation.

Conclusions

In conclusion, it could be recommend to place a maximum of 15g/30ml 95% ethanol into a tube, for either single or multiple specimens, even from different tissues, but the size must be retained constant in each experiment. It has been confirmed that this new simple and rapid method, very efficacious in separating systematic factors, offers also a good level of efficiency. Further studies are necessary to link this specimen to trait and treatment effects.

References

Cozzolino, D. and I. Murray, 2004. Identification of animal meat muscles by visible and near infrared reflectance spectroscopy. Lebensmittel Wissenschaft Und Technologie, Volume 37, 447-452.

Hildrum, K.I., T. Isaksson, T. Naes, B.N. Nilsen, M. Rodbotten and P. Lea, 1995. Near-infrared reflectance spectroscopy in the prediction of sensory properties of beef. Journal of Near Infrared Spectroscopy, Volume 3, 81-87.

Iesurum, A, T. Balbi, D. Vasapollo, A. Cicognani and C. Ghimenton, 2006. Microwave processing and ethanol-based fixation in forensic pathology. American Journal of Forensic Medical Pathology, Volume 27, 178-182

Masoero, G., A. Brugiapaglia, G. Bergoglio and R. Chicco, 2004. FT-NIR spectroscopy of fresh and treated muscle tissue in young female rabbits. 8th World Rabbit Congress, Puebla, Mexico, 1409-1415.

Masoero, G., G. Bergoglio, F. Vincenti, G. Destefanis and A. Brugiapaglia, 2005. A rapid new ethanol preparation for early muscle sampling submitted to NIR analysis: preliminary results for cattle and buffalo. Italian Journal of Animal Science, Volume 4 (Suppl. 2), 299.

Shackelford, S.D., T.L. Wheeler and M. Koohmaraie, 2004. Development of optimal protocol for visible and near-infrared reflectance spectroscopic evaluation of meat quality. Meat Science, Volume 68, 371-381.

Gene expression associated with beef sensory qualities

Carine Bernard[1], I. Cassar-Malek[1], H. Dubroeucq[1], G. Renand[2] and J.F. Hocquette[1]
[1]INRA, Unité de Recherches sur les Herbivores, UR1213, Centre de Clermont-Ferrand/Theix, F-63122 Saint-Genès-Champanelle, France
[2]INRA, Station de Génétique Quantitative et Appliquée, UR337, Domaine de Vilvert, F-78352 Jouy-en-Josas, France

Abstract

Beef quality and more particularly its sensory traits (tenderness, flavour, juiciness and colour) depend on many factors including muscle characteristics, which explain, however, less than 35% of the variability in sensory qualities. They are under the control of many genes. So to get a better understanding of the physiological processes influencing meat quality, a transcriptomic analysis was conducted. *Longissimus thoracis* muscles from fourteen Charolais bull calves were analysed using microarray technology. Gene expression profiles were compared between high and low meat quality scores of tenderness, flavour and juiciness. Several genes (n=215) were differentially expressed (Fold Change>1.4; FDR<1%). They are mostly involved in different metabolic pathways and their regulation including nucleic acid and protein metabolism, developmental processes, signal transduction, cell structure and muscle contraction. Among these genes, 33 were up-regulated in the most tender, juiciest and tastiest meats and 18 of them were highly correlated with both meat flavour and juiciness. They explained a major part of these two sensory traits variabilities. Conversely, 9 genes were up-regulated in the toughest, driest and the least tasty meats, but only one had a strong negative correlation with tenderness and explained a major part of its variability. In conclusion, the expression of some genes was associated with the sensory qualities of beef, accounting for a higher percentage of variability than the phenotypic characteristics.

Keywords: gene expression, microarray, muscle, bovine, sensory quality

Introduction

The control of beef sensory quality traits (tenderness, flavour, juiciness and colour) is important for beef producers and retailers in order to satisfy the consumer's requirements. Meat quality traits depend on both *in vivo* factors and technical factors (slaughtering conditions, ageing time, cooking process; Monson *et al.*, 2005). To date, the *in vivo* determinants of meat eating quality are not yet fully understood, but it is clear that the intrinsic characteristics of the muscle play an important role. They include proteases, collagen content, fat and fatty acid characteristics and muscle fibre type of the animal (Geay *et al.*, 2001). However, muscle fibres, collagen and lipid characteristics explained one quarter to one third only of the variability of 2 day mechanical strength and 15 day tenderness or flavour scores (Renand *et al.*, 2001). Additional muscle characteristics may therefore contribute to the variability in tenderness and flavour.

Analysis of gene expression in muscles is expected to bring new knowledge of the muscle physiological processes and their influence on meat quality. Nowadays, functional genomics provides new opportunities for determining the molecular processes related to meat quality (Eggen and Hocquette, 2004). Microarray technology represents a powerful tool for the identification of genes associated with meat sensory attributes. Only few studies have been conducted to this end either in pigs (Plastow *et al.*, 2005) or in cattle (Sudre *et al.*, 2005; Wang *et al.*, 2005), mainly due to the lack of appropriate and specific tools in livestock species. Recently, cDNA arrays have been constructed for bovine studies (for instance, see Bernard *et al.,* 2005; Lehnert *et al.,* 2004). In parallel, oligonucleotide arrays have been developed, particularly an accurate set of muscle-specific probes selected from mammal studies (Lamirault *et al.,* 2006). In this study, we have explored muscle gene expression in order to identify specific and reliable indicators of meat quality, which are differentially expressed between beef meat cuts with variable tenderness, juiciness and flavour. Long-oligonucleotide microarrays representing genes involved in skeletal and cardiac muscle physiology were used to compare gene expression profiles in *Longissimus thoracis* muscles of bull calves according to their meat quality.

Material and methods

Animals and samples

This study was conducted with 25 young Charolais bull calves from an INRA experimental herd, weaned at 32 weeks and then kept in an open shed. They were fed a complete pelleted diet distributed *ad libitum* with a limited amount of straw until slaughter. Thirteen animals were slaughtered at 15 months of age and 12 at 19 months of age. The warm carcass and the internal fat deposit weights were recorded. Carcass composition (muscle and fat contents) was estimated from dissection of the 6th rib using the Robelin and Geay (1975) prediction equation. *Longissimus thoracis* muscle (red and oxidative muscle, LT) was excised from each animal within less than ten minutes after slaughter. For biochemical and transcriptomic studies, the samples were snap-frozen in liquid nitrogen and stored at -80 °C until analysed.

Biochemical and histochemical studies

Enzyme activities, collagen characteristics, fibre areas, total lipids, triglycerides and phospholipids content, were determined as previously described (Sudre *et al.,* 2005; Cassar-Malek *et al.,* 2004).

Meat quality

LT steaks were vacuum-packaged, aged at 4 °C for 14 days post-mortem, and frozen. Steaks were thawed rapidly under flowing water. The following day, the steaks were grilled to a core temperature of 55 °C and immediately served to panelists. A total of 10 – 12 trained panelists was used in each test session. They evaluated eight meat samples, presented sequentially in each session. They scored initial tenderness and overall tenderness, juiciness and flavour intensity on 10-point scales: from 1 (tough; dry; less tasty) to 10 (tender; juicy; tasty). Meat samples from different bull calves of the same slaughter age group were randomly presented to the panelists. Each sample was tested once independently by each panelist. Scores were averaged over the different panelists for each calf. Meat texture was evaluated by Shear Force measurement with a Warner-Brätzler device. Animals were classified according to the score attributed to their meat for each sensory trait (tenderness (T), juiciness (J) and flavour (F)). For each criterion, eight animals (4 per age group) were chosen among extremes of the distribution to perform comparisons between high (+) and low (-) meat quality scores of tenderness, flavour and juiciness. A total of 14 animals extreme for tenderness, flavour and/or juiciness scores of their meat were kept for further analyses.

Transcriptomics

Total RNA was extracted from LT and its integrity was assessed using the Agilent 2100 bioanalyser. Reverse-transcription and indirect labelling were performed from 4 x 15 μg of total RNA from each sample using the CyScribe cDNA Post Labelling Kit (Amersham Pharmacia Biotech) as described by the manufacturer. Samples from each animal were labelled individually with Cy3 and mixed with an equal amount of a reference pool (LT muscle transcripts isolated from 19-month-old animals) labelled with Cy5 prior to their hybridisation onto the Myochips (50-mer oligonucleotides probes, 5418 spots in triplicate). These microarrays (75% human and 25% murine sequences representative of cardiac and skeletal genes) were prepared at the West Genopole transcriptome platform (Nantes, France; http://cardioserve.nantes.inserm.fr/ptf-puce/myochips_fr.php). After washing, hybridised arrays (n=4 per animal) were scanned by fluorescence confocal microscopy at 10 μm/pixel resolution and images were analysed using the Genepix Pro 6.0 software. Expression data were normalised using MADSCAN (LeMeur *et al.*, 2004) and differential analysis was performed with SAM (Tusher *et al.*, 2001). Differential gene expression was checked by real-time PCR.

Results

Meat sensory traits and muscle characteristics

The correlation coefficients between the 11 measured muscle traits and the sensory scores computed with muscles of 14 animals are low to moderate. Tenderness variability was predominantly related to ICDH and COX activities (oxidative metabolism) and to Warner-Brätzler shear force ($r = 0.50$, 0.37 and -0.36 respectively although non significant). Juiciness score tended to be explained by total lipid and triglyceride contents and COX activity ($r = -050$, -0.45 and +0.30 respectively although non significant). The latter as well as collagen content appeared to be correlated with flavour ($r = +0.58^*$ and +0.31 respectively, $^*P < 0.05$). Each characteristic, alone, explained less than 33% of flavour variability and less than 25% of tenderness and juiciness variability. Together, these characteristics accounted for up to 50% of sensory trait variability.

Gene expression profiles associated with meat sensory quality

Analysis of hybridisation signals revealed 4497 reliable signals over the 5418 oligonucleotide sequences, corresponding to 83% valid expression signals. A low technical variability was observed between four different arrays for each sample (CVmean = 11.3%) from valid hybridisation signals showing a good reproducibility of the experimentation. SAM analysis according to tenderness scores (T+ vs. T-) meat juiciness scores (J+ vs. J-) and flavour scores (F+ vs. F-) identified 615, 1005 and 799 genes differentially expressed genes respectively. As a whole, 1772 genes were differentially expressed and associated with tenderness and/or juiciness and/or flavour, some of them being common to at least 2 sensory traits. Two hundred fifteen of the 1772 genes were up-regulated (Fold change>1.4) or down-regulated (Fold change<0.71) in at least one condition. The 215 genes were represented by 159 human and 56 murine oligonucleotides. However, only some of them have been yet identified in the bovine genome (37% known genes and 52% EST sequences). One hundred eighty five of the oligonucleotides had a sequence homology with the bovine genome and among them, 84% had a high homology (>80%). They were mostly involved in different metabolic pathways and their regulation including nucleic acid and protein metabolism, developmental processes, signal transduction, cell structure and muscle contraction, as predicted by the Gene Ontology information (GO).

Interestingly, of the 215 genes, 29, 21 and 15 genes were differentially expressed with FC>1.4 and specific of tenderness, juiciness and flavour respectively. Moreover, 10 genes were simultaneously associated to tenderness and juiciness, 33 to flavour and juiciness and 49 to tenderness and flavour. Finally, only 58 genes among the 215 were differentially expressed according to tenderness, juiciness and flavour. These 58 genes were distributed as follows: 33 were up-regulated in muscles whose meats were the most tender, juiciest and tastiest. Nine were down-regulated in the same samples and 16 had different expression profiles between tenderness, juiciness and flavour. Ten of the 33 up-regulated genes appeared to be specific to the 15-month-old animals whereas the remaining 23 were differentially expressed at both 15 and 19 months of age.

Of the 23 differential genes, 19 belonged to a same expression cluster (data not shown). Eighteen of these were highly correlated with both meat juiciness and flavour and they explained up to 50% and 60% of juiciness and flavour variability respectively. Five of the 9 down-regulated genes had the highest differential expression ratio for all quality traits.

One down-regulated gene (DNAJA1) had a strong negative correlation with tenderness and explained up to 64% of its variability (Figure 1). The differential expression of the DNAJA1 gene was checked by real-time PCR (Genomic marker for meat tenderness. Patent EP06300943.5, September, 12th, 2006) (data not shown).

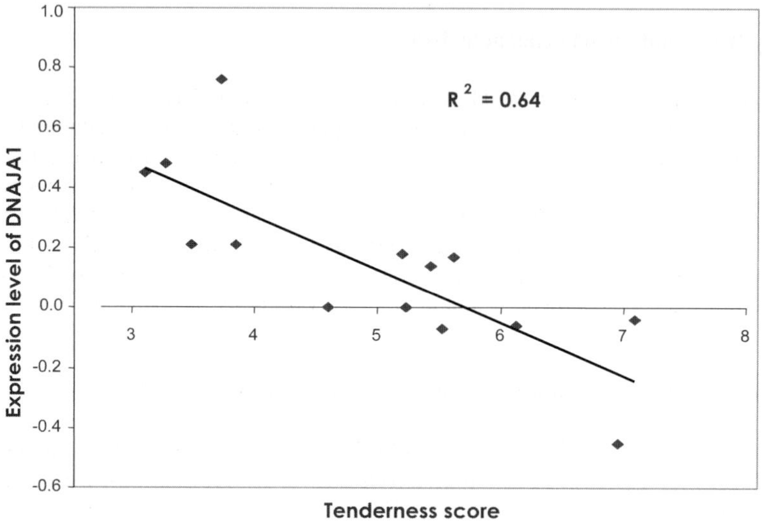

Figure 1. Expression level of DNAJA1 in Longissimus thoracis muscle of 14 Charolais young bulls according to tenderness score of their meat after 14 days of ageing.

Discussion

The study was designed to find genes whose expression was altered in *Longissimus thoracis* muscles of Charolais bull calves giving meats of different tenderness, juiciness and flavour. To this end, differences in gene expression profiles were analysed between high and low meat quality scores of tenderness, flavour and juiciness using microarray technology. The originality of this work is the identification of novel putative biological indicators of beef sensory qualities using a combination of transcriptomics and muscle and meat characteristics. Only 12% of differential genes (n=215) seemed to be interesting for sensory quality and particularly 42 genes which were differentially expressed according to the three criteria (tenderness, juiciness and flavour). Thirty-three of them appeared to be associated with high sensory quality. However, neither main biochemical pathway, nor biological process was identified as being highly involved in beef quality. This supported the hypothesis that beef quality attributes result from a combination of different biological processes (for instance, toughness of connective tissue, meat proteolysis, etc for tenderness; Maltin *et al.*, 2003).

In one hand, the experiment allowed detection of a specific indicator of tenderness. Indeed, among the five highest down-regulated genes for the three sensory traits, only one was strongly related to tenderness. This gene called DNAJA1 encodes a member of the large heat shock 40 kDa protein family. This protein is a co-chaperone of Hsc70 and seems to play a role in protein import into mitochondria. However, to date, few data are available on the relation between expression of this gene and muscle physiology, and subsequently meat quality. On the other hand, putative specific indicators of flavour and juiciness were also detected from the category of up-regulated genes highly correlated to juiciness and flavour scores. In particular, the PRKAG1 gene, which encodes a muscle-specific isoform of the protein kinase adenosine monophosphate-activated regulatory γ1−subunit, has GO information for fatty acid metabolism. Characterization of bovine PRKAG1 gene has just begun (Benkel *et al.,* 2005) and its role in beef quality is still unknown. However, mutation in the PRKAG3 gene in pigs accounts for the 'Rendement Napole' (RN) or 'Hampshire effect' gene (Milan *et al.*, 2000). It cause low ultimate pH and poor water holding capacity in pork, reduced yield of cured cooked ham, and is detrimental for meat quality. Another interesting finding was the differential expression of genes involved in iron metabolism, e.g. Ireb2 a post translational regulator of ferritin. Lastly, data mining of literature indicated that some of these genes were regulated by different transcription factors, involved notably in regulation of gene expression especially of adipocyte-specific genes, e.g. the androgen receptor, Sp1 or USF2.

Conclusion

In conclusion, our data underline the usefulness of genomics in farm animals. Moreover, they allowed the identification of differentially expressed genes associated with variability of beef tenderness, juiciness and flavour and which may be new indicators of beef quality. However, further studies are needed to get a better understanding of their biological functions in cattle and their putative involvement in meat quality.

References

Benkel, B., S. Kollers, R. Fries, A. Sazanov, E. Yoshida, E. Valle, J. Davoren and Hickey D., 2005. Characterization of the bovine ampkgamma1 gene. Mammalian Genome, Volume 16, 194-200.

Bernard, C., S. Degrelle, S. Ollier, E. Campion, I. Cassar-Malek, G. Charpigny, S. Dhorne-Pollet, I. Hue, J.F. Hocquette, F. Le Provost, C. Leroux, F. Piumi, G. Rolland, S. Uzbekova, E. Zalachas and P. Martin, 2005. A cDNA macro-array resource for gene expression profiling in ruminant tissues involved in reproduction and production (milk and beef) traits. Journal of Physiology and Pharmacology, Volume 56 (Suppl 3), 215-224.

Cassar-Malek, I., J.F. Hocquette, C. Jurie, A. Listrat, R. Jailler, D. Bauchart, Y. Briand and B. Picard, 2004. Muscle-specific metabolic, histochemical and biochemical responses to nutritionally-induced discontinuous growth path. Animal Science, Volume 79, 49-59.

Eggen, A. and J.F. Hocquette, 2004. Genomic approaches to economic trait loci and tissue expression profiling: application to muscle biochemistry and beef quality. Meat Science, Volume 66, 1-9.

Geay, Y., D. Bauchart, J.F. Hocquette and J. Culioli, 2001. Effect of nutritional factors on biochemical, structural and metabolic characteristics of muscles in ruminants, consequences on dietetic value and sensorial qualities of meat. Reproduction Nutrition Development, Volume 41, 1-26 and Erratum, 377.

Lamirault, G., N. Gaborit, N. Le Meur, C. Chevalier, G. Lande, S. Demolombe, D. Escande, S. Nattel, J.J. Léger and M. Steenman, 2006. Gene expression profile associated with chronic atrial fibrillation and underlying valvular heart disease in man. Journal of Molecular and Cellular Cardiology, Volume 40, 173-184.

LeMeur, N., G. Lamirault, A. Bihouee, M. Steenman, H. Bedrine-Ferran, R. Teusan, G. Ramstein and J. Léger, 2004. A dynamic, web-accessible resource to process raw microarray scan data into consolidated gene expression values: importance of replication. Nucleic Acids Research, Volume 32, 5349-5358.

Lehnert, S.A., Y.H. Wang and K.A. Byrne, 2004. Development and application of a bovine cDNA microarray for expression profiling of muscle and adipose tissue. Austrian Journal of Experimental Agriculture, Volume 44, 1127-1133.

Maltin, C., D. Balcerzak, R. Tilley and M. Delday, 2003. Determinants of meat quality: tenderness. Proceeding of the Nutrition Society, Volume 62, 337-347.

Milan, D., J.T. Jeon, C. Looft, V. Amarger, A. Robic, M. Thelander, C. Rogel-Gaillard, S. Paul, N. Iannuccelli, L. Rask, H. Ronne, K. Lundstrom, N. Reinsch, J. Gellin, E. Kalm, P.L. Roy, P. Chardon and L. Andersson, 2000. A mutation in PRKAG3 associated with excess glycogen content in pig skeletal muscle. Science, Volume 288, 1248-1251.

Monson, F., C. Sañudo and I. Sierra, 2005. Influence of breed and ageing time on the sensory meat quality and consumer acceptability in intensively reared beef. Meat Science, Volume71, 471-479.

Plastow, G.S., D. Carrion, M. Gil, J.A. Garcia-Regueiro, M.F.I. Furnols, M. Gispert, M.A. Oliver, A. Velarde, M.D. Guardia and M. Hortos 2005. Quality pork genes and meat production. Meat Science, Volume 3, 409-421.

Renand, G., B. Picard, C. Touraille, P. Berge and J. Lepetit, 2001. Relationships between muscle characteristics and meat quality traits of young Charolais bulls. Meat Science, Volume 59, 49-60.

Robelin, J. and Y. Geay, 1975. Estimation de la composition des carcasses de jeunes bovins à partir de la composition d'un morceau monocostal prélevé au niveau de la 11eme côte. Annales de Zootechnie, Volume 24, 391-402.

Sudre, K., I. Cassar-Malek, A. Listrat, Y. Ueda, C. Leroux, C. Jurie, C. Auffray, G. Renand, P. Martin and J.F. Hocquette, 2005. Biochemical and transcriptomic analyses of two bovine skeletal muscles in Charolais bulls divergently selected for muscle growth. Meat Science, Volume 70, 267-277.

Tusher, V.G., R. Tibshirani and G. Chu, 2001. Significance analysis of microarrays applied to the ionizing radiation response. Proceeding National Academy of Science of the United States of America, Volume 98, 5116-5121.

Wang, Y.H., K.A. Byrne, A. Reverter, G.S. Harper, M. Taniguchi, S.M. McWilliam, H. Mannen, K. Oyama and S.A.. Lehnert, 2005. Transcriptional profiling of skeletal muscle tissue from two breeds of cattle. Mammalian Genome, Volume 16, 201-210.

Author index